sona nasin pi nanpa Okitonon

Philosophy of Octonions

Philosophie der Oktonionen

Martin Erik Horn
jan Masin Elki pi ilo kalama pi palisa lawa
Lilienthalpark / ma kasi pi jan tawa sewi nanpa wan
Berlin / ma tomo Pelin
Deutschland / Germany / ma Tosi

Email:
toki-pona@clifford-algebra.de

sona nasin pi nanpa Okitonon
Philosophy of Octonions
Philosophie der Oktonionen

kepeken toki tu wan:
toki Pona, toki Inli, toki Tosi

trilingual version:
Toki Pona, English, German

dreisprachige Fassung:
Toki Pona, Englisch, Deutsch

Martin Erik Horn – jan Masin Elki
pi ilo kalama pi palisa lawa

Der Verwendung dieses Textes zum Trainieren von Sprachverar-
beitungssystemen oder KI-Tools aller Art wird nicht zugestimmt.
The autor does not allow to use this text to train language processing
systems or AI tools of any kind.

Bibliographische Information der Deutschen Nationalbibliothek:

Die Deutsche Nationalbibliothek verzeichnet diese Publikation
in der Deutschen Nationalbiographie;
detaillierte bibliographische Daten sind im Internet über

http://dnb.dnb.de

abrufbar.

© 2025 Martin Erik Horn

Verlag:
BoD · Books on Demand GmbH, Überseering 33, 22297 Hamburg,
bod@bod.de
Druck:
Libri Plureos GmbH, Friedensallee 273, 22763 Hamburg

ISBN: 978-3-7693-1156-3

pali li pana e sona.

Actions give knowledge.
One learns by experience.
Tun bringt Wissen.
Durch Erfahrung lernt man.

Sonja Lang: Toki Pona. The Language of Good.
The official Toki Pona book pu.

nanpa pi ijo nasa luka tu li pana e sona.
sona nanpa Okitonon li pana e sona.

The numbers of seven imaginary quantities give knowledge.
The algebra of octonions gives knowledge.
Die Zahlen der sieben imaginären Größen bringen Wissen.
Die Oktonionenalgebra bringt Wissen.

kipisi nanpa wan:
sona nasin pi nasin ma
luka tu

First Part:
The Philosophy of Seven
Dimensions

Erster Teil:
Die Philosophie der sieben
Dimensionen

sona nasin li suli.

Philosophy is important.
Philosophie ist wichtig.

sona tawa li suli.

Physics is important.
Physik ist wichtig.

tan ni la sona nasin pi sona tawa li suli.

Therefore the philosophy of physics is important.
Deshalb ist die Philosophie der Physik wichtig.

sona nasin ante tu li lon, lon ma ni.

On earth two different philosophies exist.
Auf der Erde gibt es zwei verschiedene Philosophien.

sona nasin ni li utala wawa.

These philosophies contradict each other.
Diese Philosophien widersprechen sich.

sona nasin nanpa wan:

First philosophy:

Erste Philosophie:

ma mi li jo e nasin ma mute.

Our world has many dimensions.
Unsere Welt besitzt viele Dimensionen.

nasin ma tu tu li lili.

Four dimensions are insufficient.
Vier Dimensionen sind zu wenig.

nasin ma mute mute li wile.

Many more dimensions are necessary.
Viel mehr Dimensionen sind notwendig.

ken ala la mi sona e ma mi, lon nasin ma tu tu taso.

It is not possible that we understand our world with four dimensions only.
Es ist nicht möglich, dass wir mit nur vier Dimensionen unsere Welt verstehen.

**nasin ma tu tu taso li lon la
mi kama jo ala e lawa pona pi sona tawa.**

If only four dimensions exist,
we will not find the laws of physics.
Wenn nur vier Dimensionen existieren,
werden wir die Gesetze der Physik nicht finden.

tan ni la jan Kaku li toki:

Therefore Michio Kaku says:
Deshalb sagt Michio Kaku:

> … a scientific revolution created
> by the theory of hyperspace
> which states that dimensions
> exist beyond the commonly
> accepted four of space and time
> … a conventional four-dimensional
> theory is 'too small' to describe
> our universe.

o lukin e ni / see / siehe:

Michio Kaku: Hyperspace. A Scientific Odyssey
Through Parallel Universes, Time Warps, and the Tenth
Dimension. Oxford University Press, Oxford 1999.

jan pi sona nanpa li pali e sona nanpa sin.

Mathematicians invent new mathematics.
Mathematiker erfinden neue Mathematiken.

**tenpo pini la jan pi sona nanpa li pali
e sona nanpa sin.**

Mathematicians have invented new mathematics.
Mathematiker haben neue Mathematiken erfunden.

sona nanpa sin ni li jo e nasin ma mute mute.

These new mathematics possess many more
dimensions.
Diese neuen mathematischen Ansätze besitzen viele
weitere Dimensionen.

**sijelo sona nanpa suli ni li jo e nasin ma mute:
sona nanpa Okitonon**

This important mathematical structure possesses
many dimensions: Octonion algebra
Diese wichtige mathematische Struktur besitzt viele
Dimensionen: Oktonionenalgebra

sona nanpa Okitonon li jo e ijo nasa pona luka tu:

Octonions have seven imaginary base units:
Oktonionen besitzen sieben imaginäre Basiseinheiten:

$$e_1, \ e_2, \ e_3, \ e_4, \ e_5, \ e_6, \ e_7$$

leko Okitonon ona li sama e wan jasima.

Their octonionic squares are identical to minus one.
Ihre oktonioninschen Quadrate sind gleich minus Eins.

$$e_1{}^2 = e_2{}^2 = e_4{}^2 = e_7{}^2 = -1$$

sona nanpa Okitonon li jo e ijo lon pona wan:

The algebra of octonions has one real base unit:
Die Oktonionenalgebra besitzt eine reelle Basiseinheit:

$$e_0 = 1$$

leko ona li sama e wan pona.

Its square is identical to plus one.
Ihr Quadrat ist gleich plus Eins.

$$e_0{}^2 = 1^2 = +1$$

ken la nanpa Okitonon li kulupu.

Octonions can be multiplied.
Oktonionen können multipliziert werden.

ona li kulupu, insa sike.

They multiply in a circular way.
Sie multiplizieren sich kreisförmig.

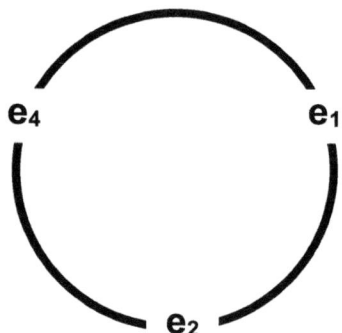

taso nasin li suli.

But the direction is important.
Aber die Richtung ist wichtig.

nasin pi ilo tenpo li lon la kulupu li pona.

Multiplication is positive in a clockwise orientation.
Im Uhrzeigersinn ist die Multiplikation positiv.

12

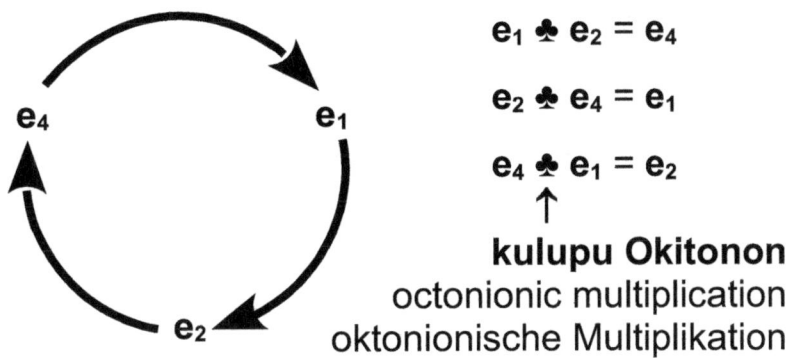

$$e_1 \clubsuit e_2 = e_4$$

$$e_2 \clubsuit e_4 = e_1$$

$$e_4 \clubsuit e_1 = e_2$$

↑

kulupu Okitonon
octonionic multiplication
oktonionische Multiplikation

ni li kulupu pi nanpa Okitonon.

This is an octionionic multiplication.
Dies ist eine Oktonionen-Multiplikation.

mi sitelen e kulupu ni pi nanpa Okitonon, kepeken sitelen ilo utala pi lipu musi.

We write this multiplication of octonions with the club symbol of playing cards.

Wir schreiben diese Multiplikation von Oktonionen mit Hilfe des Kreuz-Symbols von Spielkarten.

kulupu ni pi nanpa Okitonon li ante.

This multiplication of octonions ist different.
Diese Oktonionen-Multiplikation ist anders.

nasin jasima pi ilo tenpo li lon la kulupu li jasima.

Multiplication is negative in an anti-clockwise orientation.
Entgegen dem Uhrzeigersinn ist die Multiplikation negativ.

$$e_2 \clubsuit e_1 = - e_4$$

$$e_4 \clubsuit e_2 = - e_1$$

$$e_1 \clubsuit e_4 = - e_2$$

nanpa lon li kulupu la ante pi nasin kulupu li jo ala e sitelen jasima ante, tenpo ale.

If real numbers are multiplied, a change of the order of multiplication will be always without a change of sign.
Wenn reelle Zahlen multipliziert werden, verläuft eine Änderung der Multiplikationsreihenfolge immer ohne einem Vorzeichenwechsel.

ijo pana / examples / Beispiele:

$$3 \clubsuit 5 = 15 \qquad \longleftrightarrow \qquad 5 \clubsuit 3 = 15$$

$$3 \clubsuit (- 8) = - 24 \qquad \longleftrightarrow \qquad (- 8) \clubsuit 3 = - 24$$

$$7 \clubsuit e_2 = 7 \, e_2 \qquad \longleftrightarrow \qquad e_2 \clubsuit 7 = 7 \, e_2$$

14

nanpa nasa li kulupu la ante pi nasin kulupu li jo e sitelen jasima ante, tenpo ale.

If imaginary numbers are multiplied, a change
of the order of multiplication
will always have a change of sign.
Wenn imaginäre Zahlen multipliziert werden, verläuft
eine Änderung der Multiplikationsreihenfolge immer
mit einem Vorzeichenwechsel.

ijo pana / examples / Beispiele:

$$(3\ e_4) \clubsuit (5\ e_1) = 15\ e_2 \quad \longleftrightarrow \quad (5\ e_1) \clubsuit (3\ e_4) = -15\ e_2$$

$$(3\ e_1) \clubsuit (-8\ e_2) = -24\ e_4 \quad \longleftrightarrow \quad (-8\ e_2) \clubsuit (3\ e_1) = 24\ e_4$$

$$(7\ e_4) \clubsuit e_2 = (-7\ e_1) \quad \longleftrightarrow \quad e_2 \clubsuit (7\ e_4) = 7\ e_1$$

leko pi nanpa Okitonon li sama e kulupu pi nanpa Okitonon, kepeken ona sama.

The square of an octonion is identical to a
multiplication of an octionion by itself.
Das Quadrat eines Oktonions entspricht der
Multiplikation des Oktonions mit sich selbst.

en tan ni la mi sitelen e leko Okitonon pi ijo nasa pona Okitonon, kepeken ni:

And therefore we should write the octonionic square of an octonion base unit in the following way:
Und deshalb sollten wir das oktonionische Quadrat einer oktonionischen Basiseinheit folgendermaßen schreiben:

$$e_1 \clubsuit e_1 = e_2 \clubsuit e_2 = e_3 \clubsuit e_3 = \ldots = e_7 \clubsuit e_7 = -1$$

taso o kute e ante !

But pay attention to the difference!
Aber achte auf den Unterschied!

$$e_3^2 = e_3\, e_3 \neq e_3 \clubsuit e_3$$

$$e_5^2 = e_5\, e_5 \neq e_5 \clubsuit e_5$$

$$e_6^2 = e_6\, e_6 \neq e_6 \clubsuit e_6$$

$$\uparrow \qquad\qquad \uparrow$$

kulupu nasa ala   **kulupu Okitonon**
usual multiplication   octonionic multiplication
normale Multiplikation   Oktonionen-Multiplikation

kulupu tu ante li lon.

Two different multiplications exist.
Es existieren zwei verschiedene Multiplikationen.

16

sike ante pi sona nanpa Okitonon li mute.

There are many different octonionic circles.
Es gibt viele verschiedene oktonionische Kreise.

mi kama jo e sike nanpa tu, tenpo ni.

Now we will find the second circle.
Nun werden wir den zweiten Kreis konstruieren.

$$e_1 \clubsuit e_2 = e_4 \qquad e_2 \clubsuit e_4 = e_1 \qquad e_4 \clubsuit e_1 = e_2$$

$$\downarrow \quad \downarrow \quad \downarrow + 1 \qquad \downarrow \quad \downarrow \quad \downarrow + 1 \qquad \downarrow \quad \downarrow \quad \downarrow + 1$$

$$e_2 \clubsuit e_3 = e_5 \qquad e_3 \clubsuit e_5 = e_2 \qquad e_5 \clubsuit e_2 = e_3$$

sitelen lili ale li mute e wan.

All indices are increased by one.
Alle Indizes werden um Eins erhöht.

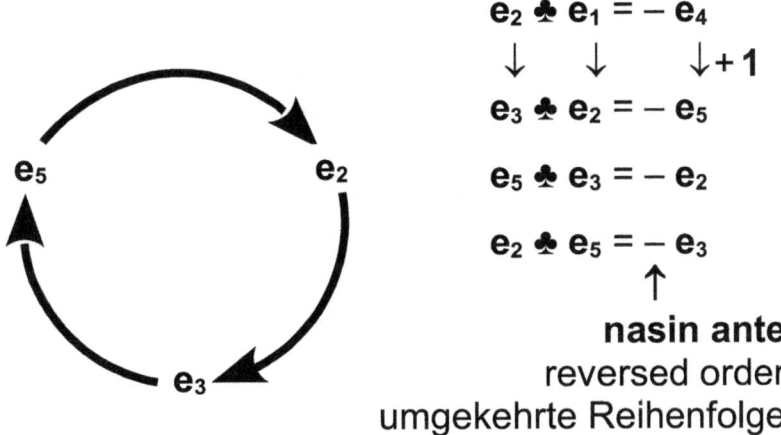

$$e_2 \clubsuit e_1 = -e_4$$

$$\downarrow \quad \downarrow \quad \downarrow + 1$$

$$e_3 \clubsuit e_2 = -e_5$$

$$e_5 \clubsuit e_3 = -e_2$$

$$e_2 \clubsuit e_5 = -e_3$$

$$\uparrow$$

nasin ante
reversed order
umgekehrte Reihenfolge

17

sike ante luka tu li lon.

Seven different circles exist.
Es gibt sieben verschiedene Kreise.

$$e_1 \clubsuit e_2 = e_4 \qquad e_2 \clubsuit e_4 = e_1 \qquad e_4 \clubsuit e_1 = e_2$$
$$\downarrow +1$$
$$e_2 \clubsuit e_3 = e_5 \qquad e_3 \clubsuit e_5 = e_2 \qquad e_5 \clubsuit e_2 = e_3$$
$$\downarrow +1$$
$$e_3 \clubsuit e_4 = e_6 \qquad e_4 \clubsuit e_6 = e_3 \qquad e_6 \clubsuit e_3 = e_4$$
$$\downarrow +1$$
$$e_4 \clubsuit e_5 = e_7 \qquad e_5 \clubsuit e_7 = e_4 \qquad e_7 \clubsuit e_4 = e_5$$
$$\downarrow -6$$
$$e_5 \clubsuit e_6 = e_1 \qquad e_6 \clubsuit e_1 = e_5 \qquad e_1 \clubsuit e_5 = e_6$$
$$\downarrow +1$$
$$e_6 \clubsuit e_7 = e_2 \qquad e_7 \clubsuit e_2 = e_6 \qquad e_2 \clubsuit e_6 = e_7$$
$$\downarrow +1$$
$$e_7 \clubsuit e_1 = e_3 \qquad e_1 \clubsuit e_3 = e_7 \qquad e_3 \clubsuit e_7 = e_1$$

sitelen lili li mute e wan.
anu ona li lili e luka wan.

Indices are increased by one.
Or they are decresed by six.
Indizes werden um eins erhöht.
Oder sie werden um sechs verringert.

en sike sike li pini.

And the circle of circles will be closed.
Und der Kreis der Kreise schließt sich.

$$e_7 \clubsuit e_1 = e_3 \qquad e_1 \clubsuit e_3 = e_7 \qquad e_3 \clubsuit e_7 = e_1$$

$$\downarrow +1 \qquad\qquad \downarrow -6$$

$$e_1 \clubsuit e_2 = e_4 \qquad e_2 \clubsuit e_4 = e_1 \qquad e_4 \clubsuit e_1 = e_2$$

jan Pano li olin ala e sike.

Fano does not like circles.
Fano mag keine Kreise.

jan Pano li olin e selo pi linja tu wan.

Fano likes triangles.
Fano mag Dreiecke.

**tan ni la Pano li sitelen e sike,
kepeken selo pi linja tu wan.**

Therefore Fano drawns the circles with triangles.
Deshalb zeichnet Fano die Kreise mit Hilfe
von Dreiecken.

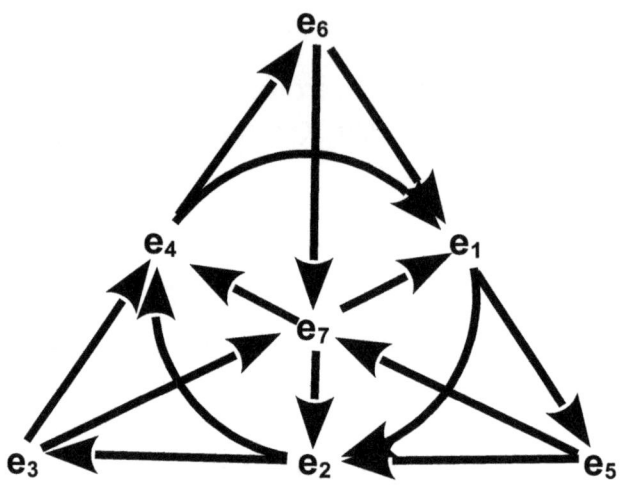

sike wan taso li awen

Only one circle survives.
Nur ein Kreis überlebt.

o lukin e ni / see / siehe:

John C. Baez: The Octonions. arXiv:math/0105155v4,
URL: https://arxiv.org/pdf/math/0105155 [23. April 2002].

nanpa Okitonon ale li sama lukin e ni:

A complete octonion then looks like:
Ein vollständiges Oktonion sieht dann so aus:

$$a = a_0 + a_1 e_1 + a_2 e_2 + a_3 e_3 + a_4 e_4 + a_5 e_5 + a_6 e_6 + a_7 e_7$$

mi kulupu, tenpo ni.

Now we multiply.
Nun multiplizieren wir.

ijo pana / example / Beispiel:

$$a = 12 + 5\,e_1$$

$$b = 3\,e_2 + 4\,e_3 \qquad \Rightarrow \qquad a \clubsuit b = ?$$

$$b \clubsuit a = ?$$

pini pona / result / Ergebnis:

$$a \clubsuit b = (12 + 5\,e_1) \clubsuit (3\,e_2 + 4\,e_3)$$

$$= 36\,e_2 + 48\,e_3 + 15\,e_1 \clubsuit e_2 + 20\,e_1 \clubsuit e_3$$

$$= 36\,e_2 + 48\,e_3 + 15\,e_4 + 20\,e_7$$

nasin ante / different order / andere Reihenfolge:

$$b \clubsuit a = (3\,e_2 + 4\,e_3) \clubsuit (12 + 5\,e_1)$$

$$= 36\,e_2 + 15\,e_2 \clubsuit e_1 + 48\,e_3 + 20\,e_3 \clubsuit e_1$$

$$= 36\,e_2 - 15\,e_4 + 48\,e_3 - 20\,e_7$$

$$= 36\,e_2 + 48\,e_3 - 15\,e_4 - 20\,e_7$$

sona pini / conclusion / Schlussfolgerung:

nasin kulupu ante la pini pona li ante.

If the order of the multiplication is changed,
the result will change.
Wenn sich die Reihenfolge der Multiplikation ändert,
wird sich das Ergebnis ändern.

$$a \clubsuit b \neq b \clubsuit a$$

pini pona li sama ala.

The results are not identical.
Die Ergebnisse sind nicht gleich.

mi wile nanpa e suli.

We will compute the magnitudes.
Wir möchten die Beträge berechnen.

tan ni la mi alasa e pali tonsi musi.

Therefore we are looking for the complex conjugation.
Deshalb suchen wir die komplexe Konjugation.

sitelen sinpin pi ijo nasa pona li ante.

The algebraic signs of the imaginary base units will change.
Die Vorzeichen der imaginären Basiseinheiten ändern sich.

$$a = a_0 + a_1 e_1 + a_2 e_2 + a_3 e_3 + a_4 e_4 + a_5 e_5 + a_6 e_6 + a_7 e_7$$
$$\downarrow \quad \downarrow \quad \downarrow \quad \downarrow \quad \downarrow \quad \downarrow \quad \downarrow$$
$$a^* = a_0 - a_1 e_1 - a_2 e_2 - a_3 e_3 - a_4 e_4 - a_5 e_5 - a_6 e_6 - a_7 e_7$$

mi sitelen e pali tonsi musi, kepeken mun lili.

We are writing the complex conjugation with a small star.
Wir schreiben die komplexe Konjugation mit Hilfe eines kleinen Sterns.

awen pi ijo pana:
Continuation of the example:
Fortsetzung des Beispiels:

$$a = 12 + 5\, e_1 \quad \Rightarrow \quad a^* = 12 - 5\, e_1$$

$$b = 3\, e_2 + 4\, e_3 \quad \Rightarrow \quad b^* = -3\, e_2 - 4\, e_3$$

mi nanpa e suli, kepeken kulupu sin.

We are computing the magnitudes with the
following multiplications.
Wir berechnen die Beträge mit Hilfe der
folgenden Multiplikationen.

$$a \clubsuit a^* = (12 + 5\,e_1) \clubsuit (12 - 5\,e_1)$$

$$= 144 - 60\,e_1 + 60\,e_1 - 25\,e_1 \clubsuit e_1$$

$$= 144 + 25$$

$$= 169$$

$$\Rightarrow \quad |a| = \sqrt{a \clubsuit a^*} = \sqrt{169} = 13$$

$$b \clubsuit b^* = (3\,e_2 + 4\,e_3) \clubsuit (-\,3\,e_2 - 4\,e_3)$$

$$= -\,9\,e_2 \clubsuit e_2 - 12\,e_2 \clubsuit e_3 - 12\,e_3 \clubsuit e_2 - 16\,e_3 \clubsuit e_3$$

$$= 9 - 12\,e_5 + 12\,e_5 + 16$$

$$= 25$$

$$\Rightarrow \quad |b| = \sqrt{b \clubsuit b^*} = \sqrt{25} = 5$$

poka: ni li lon ala e leko Okitonon.

PS: These are no octonionic squares.
PS: Das sind keine oktonionischen Quadrate.

ni li lon e leko Okitonon, lon ni:

Here are octonionic squares:
Das hier sind oktonionische Quadrate:

$$a \clubsuit a = (12 + 5 e_1) \clubsuit (12 + 5 e_1)$$
$$= 144 + 60 e_1 + 60 e_1 + 25 e_1 \clubsuit e_1$$
$$= 144 + 120 e_1 - 25$$
$$= 119 + 120 e_1$$

tenpo la leko pi nanpa Okitonon li lon ala e nanpa lon.

Sometimes squares of octonions are no real numbers.
Manchmal sind Quadrate von Oktonionen keine reellen Zahlen.

$$b \clubsuit b = (3 e_2 + 4 e_3) \clubsuit (3 e_2 + 4 e_3)$$
$$= 9 e_2 \clubsuit e_2 + 12 e_2 \clubsuit e_3 + 12 e_3 \clubsuit e_2 + 16 e_3 \clubsuit e_3$$
$$= -9 + 12 e_5 - 12 e_5 - 16$$
$$= -25$$

tenpo la leko pi nanpa Okitonon li lon e nanpa lon.

Sometimes squares of octonions are real numbers.
Manchmal sind Quadrate von Oktonionen reelle Zahlen.

mi lukin ante e suli, tenpo ni.

Now we compare the magnitudes.
Nun vergleichen wir die Beträge.

lukin ante ni li suli.

This comparison is important.
Dieser Vergleich ist wichtig.

mi lukin ante e ona, kepeken suli pi pini kulupu.

We compare them with the magnitude of the product.
Wir vergleichen sie mit dem Betrag des Produkts.

$$a \clubsuit b = 36\, e_2 + 48\, e_3 + 15\, e_4 + 20\, e_7$$

$$\Rightarrow \quad (a \clubsuit b)^* = -\,36\, e_2 - 48\, e_3 - 15\, e_4 - 20\, e_7$$

$$(a \clubsuit b) \clubsuit (a \clubsuit b)^*$$

$$= (36\, e_2 + 48\, e_3 + 15\, e_4 + 20\, e_7) \clubsuit$$
$$(-\,36\, e_2 - 48\, e_3 - 15\, e_4 - 20\, e_7)$$

$$= 36^2 + 48^2 + 15^2 + 20^2$$

$$= 1296 + 2304 + 225 + 400 = 4225$$

$$\Rightarrow \quad |a \clubsuit b| = \sqrt{(a \clubsuit b) \clubsuit (a \clubsuit b)^*} = \sqrt{4225} = 65$$

suli ni li pini kulupu pi suli tu.

This magnitude is the product of
the separate magnitudes.
Dieser Betrag entspricht dem Produkt der
einzelnen Beträge.

$$65 = 13 \clubsuit 5$$

$$\Rightarrow \quad |a \clubsuit b| = |a| \clubsuit |b|$$

lawa pona ni li suli.

This rule is important.
Diese Regel ist wichtig.

nanpa Okitonon li pali selo e sona nanpa pi suli, kepeken kipisi.

The octonioins form a normed division algebra.
Die Oktonionen bilden eine normierte
Divisionsagebra.

mi kipisi, tenpo ni.

We will divide now.
Nun werden wir teilen.

**kipisi (a ♣ b) kepeken (b) la pini pona li lon
e nanpa Okitonon (a), lon a.**

Dividing (**a ♣ b**) by **b** will result in **a**, of course.
Die Division von (**a ♣ b**) durch **b** wird natürlich
a ergeben.

mi nanpa e kipisi, kepeken pali tonsi musi.

We will compute the division by using the complex
conjugate.
Wir berechnen die Division mit Hilfe der komplexen
Konjugation.

$$|a| = \sqrt{a ♣ a^*} \qquad \Rightarrow \qquad |a|^2 = a ♣ a^*$$

$$\Rightarrow \qquad a^{-1} = \frac{a^*}{|a|^2} = \frac{a^*}{a ♣ a^*}$$

kin / furthermore / ebenfalls:

$$|b| = \sqrt{b ♣ b^*} \qquad \Rightarrow \qquad |b|^2 = b ♣ b^*$$

$$\Rightarrow \qquad b^{-1} = \frac{b^*}{|b|^2} = \frac{b^*}{b ♣ b^*}$$

nanpa / computation / Rechnung:

$$a = (a \clubsuit b) \clubsuit b^{-1} = (a \clubsuit b) \clubsuit \frac{b^*}{b \clubsuit b^*}$$

$$= (36\, e_2 + 48\, e_3 + 15\, e_4 + 20\, e_7) \clubsuit \frac{-3\, e_2 - 4\, e_3}{25}$$

$$= -\frac{1}{25}\,(108\, e_2 \clubsuit e_2 + 144\, e_2 \clubsuit e_3 + 144\, e_3 \clubsuit e_2$$

$$+ 192\, e_3 \clubsuit e_3 + 45\, e_4 \clubsuit e_2 + 60\, e_4 \clubsuit e_3$$

$$+ 60\, e_7 \clubsuit e_2 + 80\, e_7 \clubsuit e_3)$$

$$= -\frac{1}{25}\,(-108 + 144\, e_5 - 144\, e_5 - 192$$

$$- 45\, e_1 - 60\, e_6 + 60\, e_6 - 80\, e_1)$$

$$= -\frac{1}{25}\,(-300 - 125\, e_1)$$

$$= 12 + 5\, e_1$$

ni li pini pona pi awen mi.

This is the expected result.
Das ist das erwartete Ergebnis.

**taso kipisi (a ♣ b) kepeken (b) open la pini pona
li lon ala e nanpa Okitonon (a).**

But dividing (**a ♣ b**) by **b** from the left will not result in **a**.
Aber die Division von (**a ♣ b**) durch **b** von links
wird nicht **a** ergeben.

$$a \neq b^{-1} \clubsuit (a \clubsuit b) = \frac{b^*}{b \clubsuit b^*} \clubsuit (a \clubsuit b)$$

$$= \frac{-3\,e_2 - 4\,e_3}{25} \clubsuit (36\,e_2 + 48\,e_3 + 15\,e_4 + 20\,e_7)$$

$$= -\frac{1}{25}\,(108\,e_2 \clubsuit e_2 + 144\,e_2 \clubsuit e_3 + 45\,e_2 \clubsuit e_4$$

$$+ 60\,e_2 \clubsuit e_7 + 144\,e_3 \clubsuit e_2 + 192\,e_3 \clubsuit e_3$$

$$+ 60\,e_3 \clubsuit e_4 + 80\,e_3 \clubsuit e_7)$$

$$= -\frac{1}{25}\,(-108 + 144\,e_5 + 45\,e_1 - 60\,e_6$$

$$- 144\,e_5 - 192 + 60\,e_6 + 80\,e_1)$$

$$= -\frac{1}{25}\,(-300 + 125\,e_1)$$

$$= 12 - 5\,e_1$$

**pini pona ni li ike.
ni li lon ala e nanpa Okitonon (a).**

This result is challenging. This is not the octonion **a**.
Dieses Ergebnis ist herausfordernd.
Dies ist nicht das Oktonion **a**.

30

o nanpa e nanpa Okitonon (b), tenpo ni.

Let's now compute the octonion **b**.
Berechnen wir nun das Oktonion **b**.

$$b = a^{-1} \clubsuit (a \clubsuit b) = \frac{a^*}{a \clubsuit a^*} \clubsuit (a \clubsuit b)$$

$$= \frac{12 - 5\,e_1}{169} \clubsuit (36\,e_2 + 48\,e_3 + 15\,e_4 + 20\,e_7)$$

$$= \frac{1}{169} (432\,e_2 + 576\,e_3 + 180\,e_4 + 240\,e_7$$
$$- 180\,e_1 \clubsuit e_2 - 240\,e_1 \clubsuit e_3$$
$$- 75\,e_1 \clubsuit e_4 - 100\,e_1 \clubsuit e_7)$$

$$= \frac{1}{169} (432\,e_2 + 576\,e_3 + 180\,e_4 + 240\,e_7$$
$$- 180\,e_4 - 240\,e_7 + 75\,e_2 + 100\,e_3)$$

$$= \frac{1}{169} (507\,e_2 + 676\,e_3)$$

$$= 3\,e_2 + 4\,e_3$$

ni li pini pona pi awen mi.

This is the expected result.
Das ist das erwartete Ergebnis.

mi ken sitelen e nanpa ni, weka poka sike.

We are able to write these calculations
without brackets.
Wir können diese Rechnungen ohne Klammern
schreiben.

$$a = (a \clubsuit b) \clubsuit b^{-1} = a \clubsuit (b \clubsuit b^{-1}) = a \clubsuit b \clubsuit b^{-1}$$

$$a \neq b^{-1} \clubsuit (a \clubsuit b) = (b^{-1} \clubsuit a) \clubsuit b = b^{-1} \clubsuit a \clubsuit b$$

$$b = a^{-1} \clubsuit (a \clubsuit b) = (a^{-1} \clubsuit a) \clubsuit b = a^{-1} \clubsuit a \clubsuit b$$

$$b \neq (a \clubsuit b) \clubsuit a^{-1} = a \clubsuit (b \clubsuit a^{-1}) = a \clubsuit b \clubsuit a^{-1}$$

toki pi sona nanpa ni li weka poka sike.

These equations are associative.
Diese Gleichungen sind assoziativ.

toki pi nanpa Okitonon li lon ala weka poka sike, tenpo mute.

In general octonion equations are not assosicative.
Im Allgemeinen sind oktonionische Gleichungen
nicht assoziativ.

ijo pana / example / Beispiel:

$$a = 12 + 5\,e_1$$

$$b = 3\,e_2 + 4\,e_3 \qquad \Rightarrow \qquad (a \clubsuit b) \clubsuit c = ?$$

$$c = 3 + 2\,e_6 + 6\,e_7 \qquad \Rightarrow \qquad a \clubsuit (b \clubsuit c) = ?$$

pini pona nanpa wan:
First result:
Erstes Ergebnis:

$$(a \clubsuit b) \clubsuit c = ((12 + 5\,e_1) \clubsuit (3\,e_2 + 4\,e_3)) \clubsuit (3 + 2\,e_6 + 6\,e_7)$$

$$= (36\,e_2 + 48\,e_3 + 15\,e_4 + 20\,e_7) \clubsuit (3 + 2\,e_6 + 6\,e_7)$$

$$= -\,120 + 288\,e_1 + 68\,e_2 + 174\,e_3 - 51\,e_4$$
$$-\,90\,e_5 - 216\,e_6 + 132\,e_7$$

lukin pi pini pona nanpa wan:
Check of the first result:
Probe des ersten Ergebnisses:

$$a \clubsuit a^* = 12^2 + 5^2 = 169 \qquad \Rightarrow \qquad |a| = \sqrt{169} = 13$$

$$b \clubsuit b^* = 3^2 + 4^2 = 25 \qquad \Rightarrow \qquad |b| = \sqrt{25} = 5$$

$$c \clubsuit c^* = 3^2 + 2^2 + 6^2 = 49 \qquad \Rightarrow \qquad |c| = \sqrt{49} = 7$$

$$((a \clubsuit b) \clubsuit c) \clubsuit ((a \clubsuit b) \clubsuit c)^*$$

$$= (-120)^2 + 288^2 + 68^2 + 174^2 + (-51)^2$$
$$+ (-90)^2 + (-216)^2 + 132^2$$

$$= 207025$$

$$\Rightarrow \quad |(a \clubsuit b) \clubsuit c| = \sqrt{207025} = 455$$

suli ni li sama e pini kulupu pi suli wan tu.

This magnitude is identical to the product of the
three magnitudes.
Dieser Betrag entspricht dem Produkt der
drei Beträge.

$$455 = 13 \clubsuit 5 \clubsuit 7$$

$$\Rightarrow \quad |(a \clubsuit b) \clubsuit c| = |a| \clubsuit |b| \clubsuit |c|$$

lawa pona ni li suli.

This rule is important.
Diese Regel ist wichtig.

pini pona nanpa tu:
Second result:
Zweites Ergebnis:

$$a \clubsuit (b \clubsuit c) = (12 + 5\,e_1) \clubsuit ((3\,e_2 + 4\,e_3) \clubsuit (3 + 2\,e_6 + 6\,e_7))$$

$$= (12 + 5\,e_1)$$
$$\clubsuit (24\,e_1 + 9\,e_2 + 12\,e_3 - 8\,e_4 - 18\,e_6 + 6\,e_7)$$

$$= -120 + 288\,e_1 + 148\,e_2 + 114\,e_3 - 51\,e_4$$
$$+ 90\,e_5 - 216\,e_6 + 132\,e_7$$

lukin pi pini pona nanpa tu:
Check of the second result:
Probe des zweiten Ergebnisses:

$$(a \clubsuit (b \clubsuit c)) \clubsuit (a \clubsuit (b \clubsuit c))^*$$

$$= (-120)^2 + 288^2 + 148^2 + 114^2 + (-51)^2$$
$$+ 90^2 + (-216)^2 + 132^2$$

$$= 207025$$

$$\Rightarrow \quad |a \clubsuit (b \clubsuit c)| = \sqrt{207025} = 455 = 13 \clubsuit 5 \clubsuit 7$$

$$\Rightarrow \quad |a \clubsuit (b \clubsuit c)| = |a| \clubsuit |b| \clubsuit |c|$$

suli ni li lon e pini kulupu pi suli wan tu.

This magnitude is the product of the three magnitudes.
Dieser Betrag entspricht dem Produkt der drei Beträge.

sona pini / conclusion / Schlussfolgerung:

$$(a \clubsuit b) \clubsuit c \neq a \clubsuit (b \clubsuit c)$$

mi wile sitelen e poka sike, tenpo mute.

In general we have to write brackets.
Üblicherweise müssen wir Klammern setzen.

taso mi wile ala sitelen e poka sike.

But we do not want to write brackets.
Aber wir wollen keine Klammern setzen.

mi wile sitelen ala e poka sike.

We want to not write brackets.
Aber wir wollen Klammern nicht setzen müssen.

mi wile e lawa pona, weka poka sike.

Wir want rules without brackets.
Wir wollen Gesetze ohne Klammern.

**mi wile e lawa pona pi sona nanpa,
weka poka sike.**

We want mathematical rules without brackets.
Wir wollen mathematische Gesetze ohne Klammern.

mi wile e lawa pona pi sona tawa, weka poka sike.

We want rules of physics without brackets.
Wir wollen physikalische Gesetze ohne Klammern.

**ken la toki pi sona nanpa sin li sitelen,
weka poka sike:**

The following equations can be written
without brackets:
Die folgenden Gleichungen können ohne
Klammern geschrieben werden:

$$(a ♣ b) ♣ a = a ♣ (b ♣ a) = a ♣ b ♣ a$$

$$(a ♣ b) ♣ a^* = a ♣ (b ♣ a^*) = a ♣ b ♣ a^*$$

$$(a ♣ b) ♣ a^{-1} = a ♣ (b ♣ a^{-1}) = a ♣ b ♣ a^{-1}$$

**jan sona tan tomo sona Pinseton li olin
e lawa pona ni.**

Scientists from the University of Princeton like
these rule.
Wissenschaftler der Universität Princeton mögen
diese Regeln.

ona li toki / They say / Sie sagen:

"The algebra generated
by any two octonions
is associative."

o lukin e ni / see / siehe:

John H. Conway, Derek A. Smith:
On Quaternions and Octonions – Their Geometry,
Arithmetic, and Symmetry. A. K. Peters, Natick, MA 2003.
(Theorem 2, chap. 6.8, p. 76)

**sona nanpa pi nanpa Okitonon tu li lon e
sona nanpa, weka poka sike.
toki pi sona nanpa ona li weka poka sike.**

The algebra of two octonions form an algebra
without brackets.
Their equations are associative.
Eine Algebra aus zwei Oktonionen bildet eine Algebra
ohne Klammern.
Ihre Gleichungen sind assoziativ.

suli pi ijo pana / example values / Beispielwerte:

$$a = 1 + 2\,e_1 + 3\,e_2$$

$$a^* = 1 - 2\,e_1 - 3\,e_2 \qquad a \clubsuit a^* = 14$$

$$b = 4 \qquad b^* = 4 \qquad b \clubsuit b^* = 16$$

$$c = 5\,e_2 \qquad c^* = -\,5\,e_2 \qquad c \clubsuit c^* = 25$$

ijo pana nanpa wan:
First example:
Erstes Beispiel:

$$a \clubsuit b = (1 + 2\,e_1 + 3\,e_2) \clubsuit 4 = 4 + 8\,e_1 + 12\,e_2$$

$$(a \clubsuit b) \clubsuit a = (4 + 8\,e_1 + 12\,e_2) \clubsuit (1 + 2\,e_1 + 3\,e_2)$$

$$= -\,48 + 16\,e_1 + 24\,e_2$$

lukin pi pini pona / Check / Probe:

$$((a \clubsuit b) \clubsuit a) \clubsuit ((a \clubsuit b) \clubsuit a)^*$$

$$= (-48)^2 + 16^2 + 24^2$$

$$= 3136 = 14 \clubsuit 16 \clubsuit 14$$

$$= (a \clubsuit a^*) \clubsuit (b \clubsuit b^*) \clubsuit (a \clubsuit a^*)$$

$$\Rightarrow \quad \textbf{lon} \text{ / o.k.}$$

poka sike ante:
Different brackets / Andere Klammersetzung:

$$b \clubsuit a = 4 \clubsuit (1 + 2 e_1 + 3 e_2) = 4 + 8 e_1 + 12 e_2$$

$$a \clubsuit (b \clubsuit a) = (1 + 2 e_1 + 3 e_2) \clubsuit (4 + 8 e_1 + 12 e_2)$$

$$= -48 + 16 e_1 + 24 e_2$$

$$\Rightarrow \quad (a \clubsuit b) \clubsuit a = a \clubsuit (b \clubsuit a) = a \clubsuit b \clubsuit a$$

ni li pini pona pi awen mi.

This is the expected result.
Das ist das erwartete Ergebnis.

ijo pana nanpa tu:
Second example:
Zweites Beispiel:

$$a \clubsuit c = (1 + 2\,e_1 + 3\,e_2) \clubsuit (5\,e_2) = -15 + 5\,e_2 + 10\,e_4$$

$$(a \clubsuit c) \clubsuit a = (-15 + 5\,e_2 + 10\,e_4) \clubsuit (1 + 2\,e_1 + 3\,e_2)$$

$$= -30 - 60\,e_1 - 20\,e_2$$

lukin pi pini pona / Check / Probe:

$$((a \clubsuit c) \clubsuit a) \clubsuit ((a \clubsuit c) \clubsuit a)^*$$

$$= (-30)^2 + (-60)^2 + (-20)^2$$

$$= 4900 = 14 \clubsuit 25 \clubsuit 14$$

$$= (a \clubsuit a^*) \clubsuit (c \clubsuit c^*) \clubsuit (a \clubsuit a^*)$$

$$\Rightarrow \quad \textbf{lon} / \text{o.k.}$$

poka sike ante:
Different brackets / Andere Klammersetzung:

$$c \clubsuit a = (5\,e_2) \clubsuit (1 + 2\,e_1 + 3\,e_2) = -15 + 5\,e_2 - 10\,e_4$$

$$a \clubsuit (c \clubsuit a) = (1 + 2\,e_1 + 3\,e_2) \clubsuit (-15 + 5\,e_2 - 10\,e_4)$$

$$= -30 - 60\,e_1 - 20\,e_2$$

$$\Rightarrow \quad (a \clubsuit c) \clubsuit a = a \clubsuit (c \clubsuit a) = a \clubsuit c \clubsuit a$$

ni li pini pona pi awen mi.

This is the expected result.
Das ist das erwartete Ergebnis.

---

ni li pona ala.

This is not beautiful.
Das ist nicht schön.

pini pona ni li ike !

These results are ugly!
Diese Ergebnisse sind hässlich!

$$b \longrightarrow a \clubsuit b \clubsuit a$$

$$4 \longrightarrow -48 + 16\, e_1 + 24\, e_2$$

nanpa lon
real number
reelle Zahl

nanpa lon ala
no real number
keine reelle Zahl

toki pi sona nanpa li ike, tan ni: nanpa lon li kama ala e nanpa lon, tenpo sin.

The equation is ugly, because
real numbers do not become real numbers again.
Die Gleichung is hässlich, weil reelle Zahlen
nicht wieder reelle Zahlen werden.

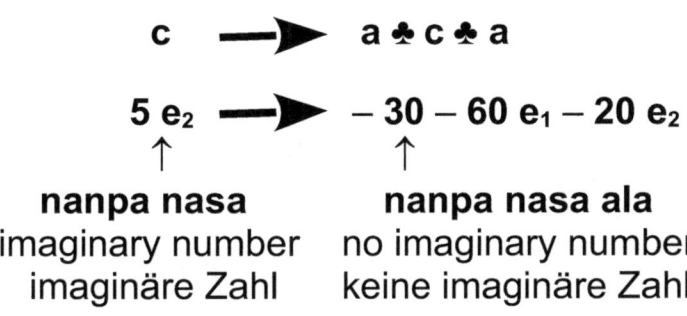

$$c \longrightarrow a \clubsuit c \clubsuit a$$

$$5\,e_2 \longrightarrow -30 - 60\,e_1 - 20\,e_2$$

nanpa nasa	nanpa nasa ala
imaginary number	no imaginary number
imaginäre Zahl	keine imaginäre Zahl

toki pi sona nanpa li ike, tan ni: nanpa nasa li kama ala e nanpa nasa, tenpo sin.

The equation is ugly, because imaginary numbers
do not become imaginary numbers again.
Die Gleichung ist hässlich, weil imaginäre Zahlen
nicht wieder imaginäre Zahlen werden.

mi wile kama jo e toki pi sona nanpa pona.

We want to get beautiful equations.
Wir möchten schöne Gleichungen bekommen.

mi wile nanpa e toki pi sona nanpa (a ♣ b ♣ a*).

We will calculate the equation **a ♣ b ♣ a***.
Wir werden die Gleichung **a ♣ b ♣ a*** berechnen.

ijo pana nanpa tu wan:
Third example:
Drittes Beispiel:

$$\overbrace{\text{a ♣ b}}$$

$$(a ♣ b) ♣ a^* = \overbrace{(4 + 8\, e_1 + 12\, e_2)}^{a ♣ b} ♣ (1 - 2\, e_1 - 3\, e_2)$$

$$= 56$$

$$a ♣ (b ♣ a^*) = (1 + 2\, e_1 + 3\, e_2) ♣ \overbrace{(4 - 8\, e_1 - 12\, e_2)}^{b ♣ a^*}$$

$$= 56$$

$$\Rightarrow \quad (a ♣ b) ♣ a^* = a ♣ (b ♣ a^*) = a ♣ b ♣ a^*$$

ni li pini pona pi awen mi.

This is the expected result.
Das ist das erwartete Ergebnis.

en toki pi sona nanpa ni li pona.

And this equation is beautiful.
Und diese Gleichung ist schön.

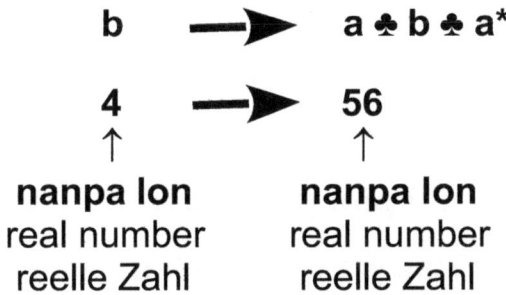

$$b \longrightarrow a \clubsuit b \clubsuit a^*$$

$$4 \longrightarrow 56$$

↑	↑
nanpa lon	**nanpa lon**
real number	real number
reelle Zahl	reelle Zahl

nanpa lon li kama lon e nanpa lon, tenpo sin.

Real numbers become real numbers again.
Reelle Zahlen werden wieder reelle Zahlen.

pana nanpa tu tu:
Fourth example:
Viertes Beispiel:

$$\overbrace{a \clubsuit c}$$
$$(a \clubsuit c) \clubsuit a^* = (-15 + 5\,e_2 + 10\,e_4) \clubsuit (1 - 2\,e_1 - 3\,e_2)$$
$$= 60\,e_1 + 30\,e_2 + 20\,e_4$$

45

$$\overbrace{}^{c \clubsuit a^*}$$

$$a \clubsuit (c \clubsuit a^*) = (1 + 2\,e_1 + 3\,e_2) \clubsuit (15 + 5\,e_2 + 10\,e_4)$$

$$= 60\,e_1 + 30\,e_2 + 20\,e_4$$

$$\Rightarrow \quad (a \clubsuit c) \clubsuit a^* = a \clubsuit (c \clubsuit a^*) = a \clubsuit c \clubsuit a^*$$

ni li pini pona pi awen mi.

This is the expected result.
Das ist das erwartete Ergebnis.

en toki pi sona nanpa ni li pona.

And this equation is beautiful.
Und diese Gleichung ist schön.

$$c \longrightarrow a \clubsuit c \clubsuit a^*$$

$$5\,e_2 \longrightarrow 60\,e_1 + 30\,e_2 + 20\,e_4$$

$$\uparrow \qquad\qquad\qquad \uparrow$$

nanpa nasa　　　　**nanpa nasa**
imaginary number　　imaginary number
imaginäre Zahl　　　　imaginäre Zahl

nanpa nasa li kama lon e nanpa nasa, tenpo sin.

Imaginary numbers become imaginary numbers again.
Imaginäre Zahlen werden wieder imaginäre Zahlen.

ni li ike ala !

This is not ugly!
Das ist nicht hässlich!

pini pona ni li pona.

These results are beautiful.
Diese Ergebnisse sind schön.

o awen sona e jan Konwe nanpa wan !

Remember Arthur W. Conway !
Denkt an Arthur W. Conway !

jan Puwe li olin e toki pi sona nanpa pona.

Furey loves beautiful equations.
Furey liebt schöne Gleichungen.

toki pi sona nanpa mute li pona.

Many equations are beautiful.
Viele Gleichungen sind schön.

toki pi sona nanpa pi nanpa Kawatenon mute en toki pi sona nanpa pi nanpa Okitonon mute li pona.

Many quaternionic equations and many octonionic
equations are beautiful.
Viele Quaternionen-Gleichungen und viele
Oktonionen-Gleichungen sind schön.

o lukin e ni / see / siehe:

Cohl Furey: How the complex quaternions give each
of the Lorentz reps of the standard model. Video 7/14
of the youTube video series „Division algebra and the
standard model [19.06.2017].
URL: https://www.youtube.com/watch?v=d3y72uw7M5Q

tan ni la jan Puwe li olin e jan Konwe nanpa wan.

Therefore Furey loves the first Conway.
Deshalb mag Furey den ersten Conway.

o lukin e ni / see / siehe:

Arthur W. Conway: Quaternion Treatment of the Relativistic Wave Equation. In: Proceedings of the Royal Society A, Vol. 162, No. 909 (1937), pp. 145 – 154.

jan Konwe nanpa wan en jan Puwe li sona e ma mi.

Arthur W. Conway and Cohl Furey understand our world.
Arthur W. Conway und Cohl Furey verstehen unsere Welt.

ona li sona e ma mi, tan ni: ona li sona e sona nanpa pona.

They understand our world, because they understand the beautiful mathematics.
Sie verstehen unsere Welt, weil sie die schöne Mathematik verstehen.

pali pana namako:
Additional problem:
Weitere Aufgabe:

$$a = 1 + 1\,e_1 + 3\,e_2 + 5\,e_3 + 8\,e_4$$

$$b = 6$$

$$c = 8\,e_4$$

$(a \clubsuit b) \clubsuit a = ?$ $a \clubsuit (b \clubsuit a) = ?$

$(a \clubsuit b) \clubsuit a^* = ?$ $a \clubsuit (b \clubsuit a^*) = ?$

$(a \clubsuit c) \clubsuit a = ?$ $a \clubsuit (c \clubsuit a) = ?$

$(a \clubsuit c) \clubsuit a^* = ?$ $a \clubsuit (c \clubsuit a^*) = ?$

o nanpa ! o lukin pi pini pona !
Compute and check the result!
Berechne und mache eine Probe!

pini pona / Solution / Lösung:

$$(a \clubsuit b) \clubsuit a = -\,588 + 12\,e_1 + 36\,e_2 + 60\,e_3 + 96\,e_4$$

$$a \clubsuit (b \clubsuit a) = -\,588 + 12\,e_1 + 36\,e_2 + 60\,e_3 + 96\,e_4$$

$\Rightarrow$ **ni li pona ala.**

$\Rightarrow \qquad (a \clubsuit b) \clubsuit a = a \clubsuit (b \clubsuit a) = a \clubsuit b \clubsuit a$

$\Rightarrow \qquad$ **toki pi sona nanpa ni li weka poka sike.**

$$(1^2 + 1^2 + 3^2 + 5^2 + 8^2) \clubsuit 6^2 \clubsuit (1^2 + 1^2 + 3^2 + 5^2 + 8^2)$$
$$= 100 \clubsuit 36 \clubsuit 100 = 360000$$

$$(-588)^2 + 12^2 + 36^2 + 60^2 + 96^2 = 360000 \qquad \Rightarrow \qquad \textbf{lon}$$

$$(a \clubsuit b) \clubsuit a^* = a \clubsuit (b \clubsuit a^*) = a \clubsuit b \clubsuit a^* = 600$$

$\Rightarrow \qquad$ **ni li pona.**

$\Rightarrow \qquad$ **toki pi sona nanpa ni li weka poka sike.**

$$600^2 = 360000 \qquad \Rightarrow \qquad \textbf{lon}$$

$$(a \clubsuit c) \clubsuit a = a \clubsuit (c \clubsuit a) = a \clubsuit c \clubsuit a$$
$$= -128 - 128\, e_1 - 384\, e_2 - 640\, e_3 - 224\, e_4$$

$\Rightarrow \qquad$ **ni li pona ala.**

$\Rightarrow \qquad$ **toki pi sona nanpa ni li weka poka sike.**

$$(-128)^2 + (-128)^2 + (-384)^2 + (-640)^2 + (-224)^2$$
$$= 640000 = 100 \clubsuit 64 \clubsuit 100 \qquad \Rightarrow \qquad \textbf{lon}$$

$$(a \clubsuit c) \clubsuit a^* = a \clubsuit (c \clubsuit a^*) = a \clubsuit c \clubsuit a^*$$
$$= 176\, e_1 + 368\, e_2 + 640\, e_3 + 240\, e_4 + 80\, e_6$$

$\Rightarrow$ ni li pona.

$\Rightarrow$ toki pi sona nanpa ni li weka poka sike.

$$176^2 + 368^2 + 640^2 + 240^2 + 80^2 = 640000 \quad \Rightarrow \quad \text{lon}$$

kipisi nanpa tu:
sona nasin pi nasin ma
tu wan
(sona nasin pi jan Pu en jan Soje)

Second Part:
The Philosophy of three Dimensions
(The Philosophy of Foo and Soy')

Zweiter Teil:
Die Philosophie der drei Dimensionen
(Die Philosophie von Foo and Soy')

sona nasin en sona tawa li suli.

Philosophy and physics are important.
Philosophie und Physik sind wichtig.

tan ni la sona nasin pi sona tawa li suli.

Therefore the philosophy of physics is important.
Deshalb ist die Philosophie der Physik wichtig.

sona nasin tu li lon, lon ma ni. ona li utala wawa.

On earth two different philosophies exist which
contradict each other.
Auf der Erde gibt es zwei verschiedene Philosophien,
die sich einander widersprechen.

sona nasin nanpa tu:

Second philosophy:
Zweite Philosophie:

ma mi li jo e nasin ma tu wan.

Our world has three dimensions.
Unsere Welt besitzt drei Dimensionen.

ona li lon e nasin ma tu wan pi ma.

They are three spatial dimensions.
Es sind drei räumliche Dimensionen.

nasin ma tu wan pi ma li lili ala.

Three spatial dimensions are not insufficient.
Drei räumlich Dimensionen sind nicht zu wenig.

ken la mi sona e ma mi, lon nasin ma tu wan taso.

It is possible that we understand our world with only
three spatial dimensions.
Es ist möglich, dass wir mit nur drei räumlichen
Dimensionen unsere Welt verstehen.

ni li sona nasin pi pona.

This is the philosophy of simplicity.
Dies ist die Philosophie der Einfachheit.

sona nasin ni li pona.

This philosophy is simple.
Diese Philosophie ist einfach.

ona li pona, tan ni: ma mi li pona.

It is simple, because our world is simple.
Sie ist einfach, weil unsere Welt einfach ist.

tan ni la jan sona tan tomo sona Kenpite li toki:

Therefore the scientists from the University of
Cambridge say:
Deshalb sagen die Wissenschaftler der
Universität Cambridge:

> … the use of higher dimensions
> (…) just seems to us to be
> unnecessary at present,
> when the algebra of the space
> that we do observe
> contains so many wonders
> that are not yet generally
> appreciated.

o lukin e ni / see / siehe:

Stephen Gull, Anthony Lasenby, Chris Doran:
Imaginary Numbers are not Real – The Geometric
Algebra of Spacetime. Foundation of Physics,
Vol. 23, No. 9 (1993), pp. 1175 – 1201.

mi pakala e ijo nasa pona luka tu.

We destroy the seven imaginary base units.
Wir zerstören die sieben imaginären Basiseinheiten.

mi ante e ona.

We replace them.
Wir ersetzen sie.

mi ante e ona, kepeken ijo weka poka sike.

We replace them by associative units.
Wir ersetzen sie durch assoziative Einheiten.

**mi ante e ona, kepeken linja nasin, kepeken tu
linja nasin, kepeken tu wan linja nasin.**

We replace them by vectors, bivectors, and
the trivector.
Wir ersetzen sie durch Vektoren, Bivektoren und
den Trivektor.

mi ante e ona, kepeken sona nanpa Paluli.

We replace them by Pauli algebra.
Wir ersetzen sie durch die Pauli-Algebra.

ni li sona nasin pi jan Pu e jan Soje.

This is the philosophy of Foo and Soy'.
Dies ist die Philosophie von Foo und Soy'.

o lukin e ni / see / siehe:

Clumsy Foo, qam Soy': QIH'batlh. Ehrenvolle Zerstörung.
Logisch Subduktive Klingonische Philosophie der
Gegenwart. BoD, Norderstedt 2021.

ma mi li pona, tan ni: ma mi li weka poka sike.

Our world is simple, because our world
is associative.
Unsere Welt einfach ist, weil unsere Welt
assoziativ ist.

mi ante e sona nanpa Okitonon, kepeken sona nanpa Paluli.

We replace octonion algebra by Pauli algebra.
Wir ersetzen die Oktonionenalgebra durch
die Pauli-Algebra.

**sona nanpa Paluli li jo e linja nasin pona tu wan
e tu linja nasin pona tu wan
e tu wan linja nasin pona wan.**

Pauli algebra consists of three base vectors, three
base bivectors, and one base trivector.
Die Pauli-Algebra besteht aus drei Basisvektoren, drei
Basis-Bivektoren und einem Basis-Trivektor.

linja nasin pona:

Base vectors: $\qquad\qquad$ $\sigma_x, \quad \sigma_y, \quad \sigma_z$
Basisvektoren:

---

lawa pona nanpa wan:
First basic rule:
Erste Grundregel:

$$\sigma_x{}^2 = \sigma_y{}^2 = \sigma_z{}^2 = 1$$

59

tu linja nasin pona:

Base bivectors: $\sigma_x\sigma_y, \quad \sigma_y\sigma_z, \quad \sigma_z\sigma_x$
Basis-Bivektoren:

---

lawa pona nanpa tu:
Second basic rule:
Zweite Grundregel:

$$\sigma_x\sigma_y = -\sigma_y\sigma_x$$

$$\sigma_y\sigma_z = -\sigma_z\sigma_y$$

$$\sigma_z\sigma_x = -\sigma_x\sigma_z$$

---

o lukin e ni / see / siehe: siehe:

Martin Erik Horn: sona nanpa Paluli – Pauli Algebra. kepeken toki tu wan: toki Pona, toki Inli, toki Tosi, trilingual version: Toki Pona, English, German, dreisprachige Fassung: Toki Pona, Englisch, Deutsch, BoD, Norderstedt 2024.

linja nasin li lon e pini mute pi linja nasin pona.

Vectors are linear combinations of base vectors.
Vektoren sind Linearkombinationen von Basisvektoren.

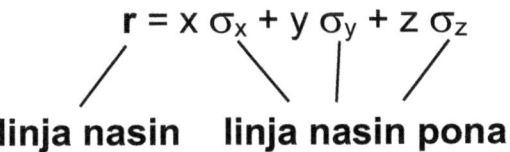

$$r = x\ \sigma_x + y\ \sigma_y + z\ \sigma_z$$

linja nasin   linja nasin pona

ijo pana / example / Beispiel:

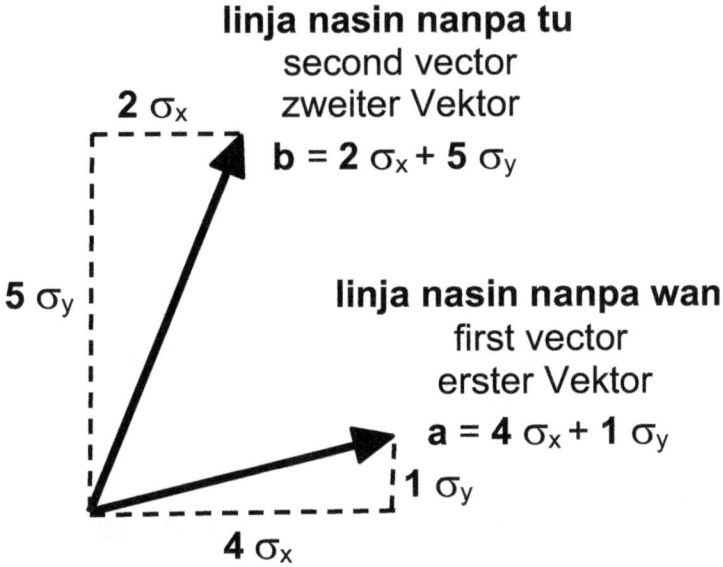

linja nasin nanpa tu
second vector
zweiter Vektor

$2\ \sigma_x$

$b = 2\ \sigma_x + 5\ \sigma_y$

$5\ \sigma_y$

linja nasin nanpa wan
first vector
erster Vektor

$a = 4\ \sigma_x + 1\ \sigma_y$

$1\ \sigma_y$

$4\ \sigma_x$

tu linja nasin li lon e pini mute pi tu linja nasin pona.

Bivectors are linear combinations of base bivectors.
Bivektoren sind Linearkombinationen von Basis-Bivektoren.

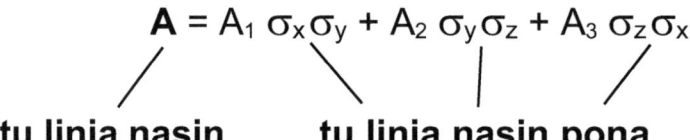

$$\mathbf{A} = A_1\, \sigma_x \sigma_y + A_2\, \sigma_y \sigma_z + A_3\, \sigma_z \sigma_x$$

tu linja nasin   **tu linja nasin pona**

(kepeken toki Inli: area … **A)**

linja nasin li kulupu la tu linja nasin li kama.

If vectors are multiplied, bivectors will be formed.
Wenn Vektoren multipliziert werden, dann entstehen Bivektoren.

taso o kute e kipisi ale !

But pay attention to all terms!
Aber achte auf alle Terme!

linja nasin li kulupu la nanpa lon li kama.

If vectors are multiplied, scalars will be formed.
Wenn Vektoren multipliziert werden, entstehen Skalare.

ijo pana / example / Beispiel:

$$P = a\,b = (4\,\sigma_x + 1\,\sigma_y)(2\,\sigma_x + 5\,\sigma_y)$$

$$= 8\,\sigma_x^2 + 20\,\sigma_x\sigma_y + 2\,\sigma_y\sigma_x + 5\,\sigma_y^2$$

$$\quad\quad\searrow\,1 \quad\quad\quad\quad \searrow -\sigma_x\sigma_y \quad\searrow\,1$$

$$= 8 + 20\,\sigma_x\sigma_y - 2\,\sigma_x\sigma_y + 5$$

$$= 13 + 18\,\sigma_x\sigma_y$$

$$\swarrow \quad\quad\quad\quad \searrow$$

nanpa lon   **tu linja nasin**

o awen sona ! sona nasin ante tu li lon.

Remember: Two different philosophies exist.
Erinnern wir uns: Es existieren zwei verschiedene
Philosophien.

tan ni la kulupu tu ante li lon, kin.

Therefore two different multiplications exist, too.
Deshalb existieren auch zwei verschiedene
Multiplikationen.

o kute e ante !

Pay attention to the difference!
Achte auf den Unterschied!

$$e_3^2 = e_3 \, e_3 \neq e_3 \clubsuit e_3$$

$$\sigma_x^2 = \sigma_x \, \sigma_x \neq \sigma_x \clubsuit \sigma_x$$

$$P = a \, b \neq a \clubsuit b$$

$$\qquad\quad \uparrow \qquad\quad \uparrow$$

kulupu nasa ala   **kulupu Okitonon**
usual multiplication   octonionic multiplication
normale Multiplikation   Oktonionen-Multiplikation

$$a \clubsuit b = (4 \, \sigma_x + 1 \, \sigma_y) \clubsuit (2 \, \sigma_x + 5 \, \sigma_y)$$

$$= 8 \, \sigma_x \clubsuit \sigma_x + 20 \, \sigma_x \clubsuit \sigma_y + 2 \, \sigma_y \clubsuit \sigma_x + 5 \, \sigma_y \clubsuit \sigma_y$$

$$= -8 + 20 \, \sigma_x \sigma_y - 2 \, \sigma_y \sigma_x - 5$$

$$= -13 + 18 \, \sigma_x \sigma_y \quad \neq \quad 13 + 18 \, \sigma_x \sigma_y = a \, b$$

ni li pali, kepeken nasin seme ?

How does this work?
Wie funktioniert das?

ni li pali, kepeken selo pi linja nasin sama.

This works with parallelograms.
Das funktioniert mit Parallelogrammen.

**linja nasin li kulupu la nanpa lon li kama.
en tu linja nasin li kama.**

If vectors are multiplied, scalars and bivectors
will be formed.
Wenn Vektoren multipliziert werden, dann bilden sich
Skalare und Bivektoren.

**nanpa lon en tu linja nasin li lon wan e selo pi linja
nasin sama.**

A scalar and a bivector together are a parallelogram.
Ein Skalar und ein Bivektor sind zusammen ein
Parallelogramm.

**linja nasin li kulupu la selo pi linja nasin sama
li kama.**

If vectors are multiplied, a parallelogram will
be formed.
Wenn Vektoren multipliziert werden, dann bildet sich
ein Parallelogramm.

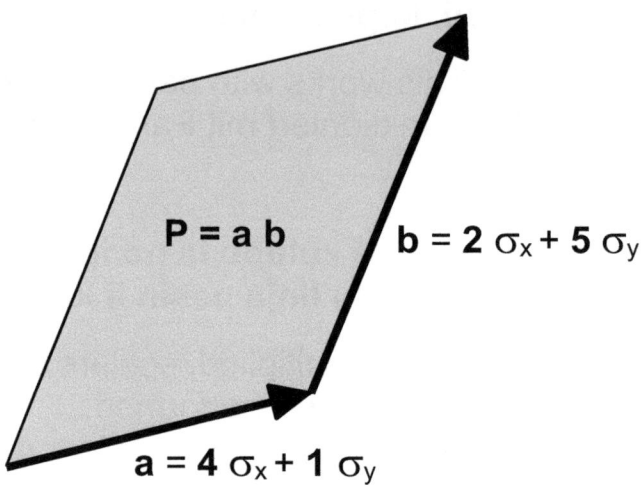

$$P = a\,b$$

$$b = 2\,\sigma_x + 5\,\sigma_y$$

$$a = 4\,\sigma_x + 1\,\sigma_y$$

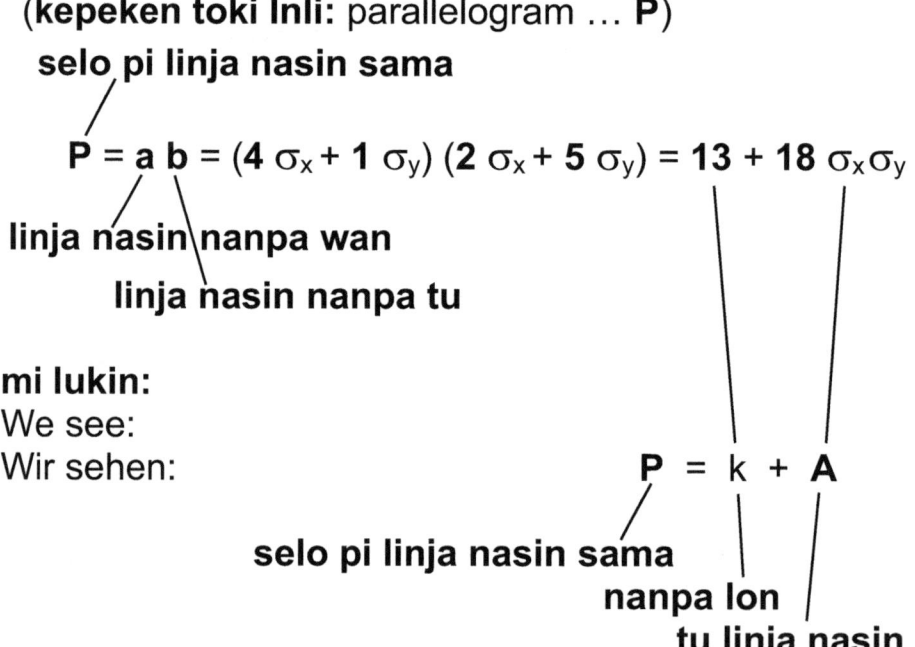

(**kepeken toki Inli:** parallelogram … **P**)

selo pi linja nasin sama

$$P = a\,b = (4\,\sigma_x + 1\,\sigma_y)(2\,\sigma_x + 5\,\sigma_y) = 13 + 18\,\sigma_x\sigma_y$$

linja nasin nanpa wan

linja nasin nanpa tu

mi lukin:
We see:
Wir sehen:

$$P = k + A$$

selo pi linja nasin sama

nanpa lon

tu linja nasin

taso o kute e ni:
selo pi linja nasin sama ante tu li lon.

But pay attention:
Two different parallelograms exist.
Aber beachte:
Es gibt zwei verschiedene Parallelogramme.

ni li selo pi linja nasin sama sin, lon ni.

Here is the new parallelogram.
Hier ist das neue Parallelogramm.

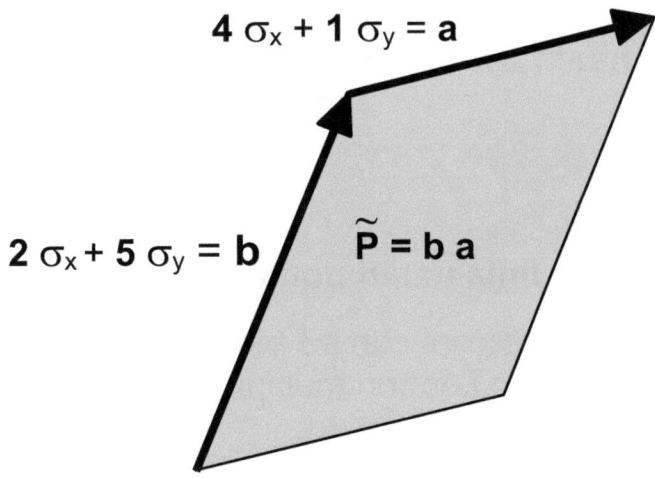

$4\,\sigma_x + 1\,\sigma_y = a$

$2\,\sigma_x + 5\,\sigma_y = b$

$\tilde{P} = b\,a$

selo pi linja nasin sama sin / new parallelogram:

neues Parallelogramm:

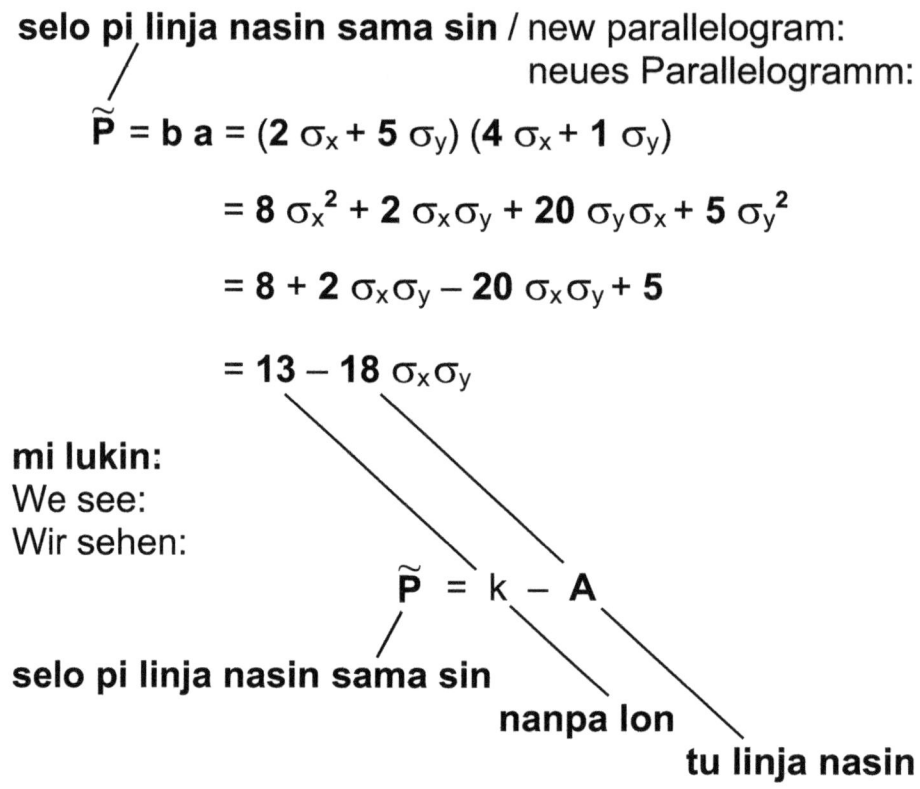

$$\tilde{P} = b\,a = (2\,\sigma_x + 5\,\sigma_y)(4\,\sigma_x + 1\,\sigma_y)$$

$$= 8\,\sigma_x{}^2 + 2\,\sigma_x\sigma_y + 20\,\sigma_y\sigma_x + 5\,\sigma_y{}^2$$

$$= 8 + 2\,\sigma_x\sigma_y - 20\,\sigma_x\sigma_y + 5$$

$$= 13 - 18\,\sigma_x\sigma_y$$

mi lukin:
We see:
Wir sehen:

$$\tilde{P} = k - A$$

selo pi linja nasin sama sin

nanpa lon

tu linja nasin

ken la tu linja nasin pona li kulupu.

Base bivectors can be multiplied.
Basis-Bivektoren können multipliziert werden.

ona li kulupu, insa sike

They multiply in a circular way.
Sie multiplizieren sich kreisförmig.

68

taso pakala suli li lon, lon ni:

But there is a big mistake here:
Aber hier gibt es einen großen Fehler:

sitelen jasima li lon e ike.
The minus sign is a problem.
Das Minuszeichen ist ein Problem.

$$\downarrow$$

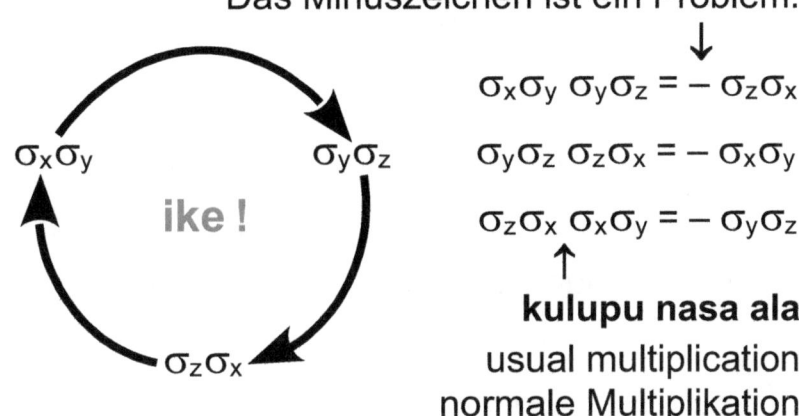

$$\sigma_x\sigma_y \; \sigma_y\sigma_z = -\,\sigma_z\sigma_x$$

$$\sigma_y\sigma_z \; \sigma_z\sigma_x = -\,\sigma_x\sigma_y$$

$$\sigma_z\sigma_x \; \sigma_x\sigma_y = -\,\sigma_y\sigma_z$$

$$\uparrow$$

kulupu nasa ala
usual multiplication
normale Multiplikation

tan ni la mi pana e sitelen jasima tu wan.

Therefore we add three minus signs.
Deshalb ergänzen wir drei Minuszeichen.

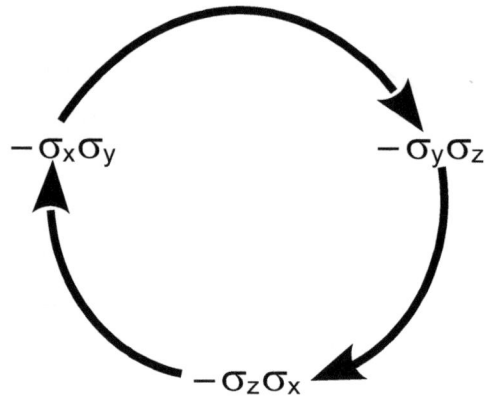

kulupu li pona, tenpo ni.

The multiplication is beautiful now.
Nun ist die Multiplikation schön.

$$(-\sigma_x\sigma_y)\,(-\sigma_y\sigma_z) = -\sigma_z\sigma_x$$

$$(-\sigma_y\sigma_z)\,(-\sigma_z\sigma_x) = -\sigma_x\sigma_y$$

$$(-\sigma_z\sigma_x)\,(-\sigma_x\sigma_y) = -\sigma_y\sigma_z$$

kulupu ni li pona mute.

This multiplication is very beautiful.
Diese Multiplikation ist sehr schön.

kulupu nasa ala en kulupu Okitonon li sama, lon ni.

Here the usual multiplication and the octonionic
multiplication are identical.
Hier sind die normale Multiplikation und die
Oktonionen-Multiplikation identisch.

$$(-\sigma_x\sigma_y)(-\sigma_y\sigma_z) = (-\sigma_x\sigma_y) \clubsuit (-\sigma_y\sigma_z) = -\sigma_z\sigma_x$$

$$(-\sigma_y\sigma_z)(-\sigma_z\sigma_x) = (-\sigma_y\sigma_z) \clubsuit (-\sigma_z\sigma_x) = -\sigma_x\sigma_y$$

$$(-\sigma_z\sigma_x)(-\sigma_x\sigma_y) = (-\sigma_z\sigma_x) \clubsuit (-\sigma_x\sigma_y) = -\sigma_y\sigma_z$$

**tan ni la tu linja nasin pona jasima pi sona nanpa
Paluli li lon e ijo Kawatenon pona.
en ona li lon e ijo Okitonon pona.**

Therefore the negative base bivectors of Pauli algebra
are quaternion base units.
And they are octonion base units.
Deshalb sind die negativen Basis-Bivektoren der
Pauli-Algebra quaternionische Basiseinheiten.
Und sie sind oktonionische Basiseinheiten.

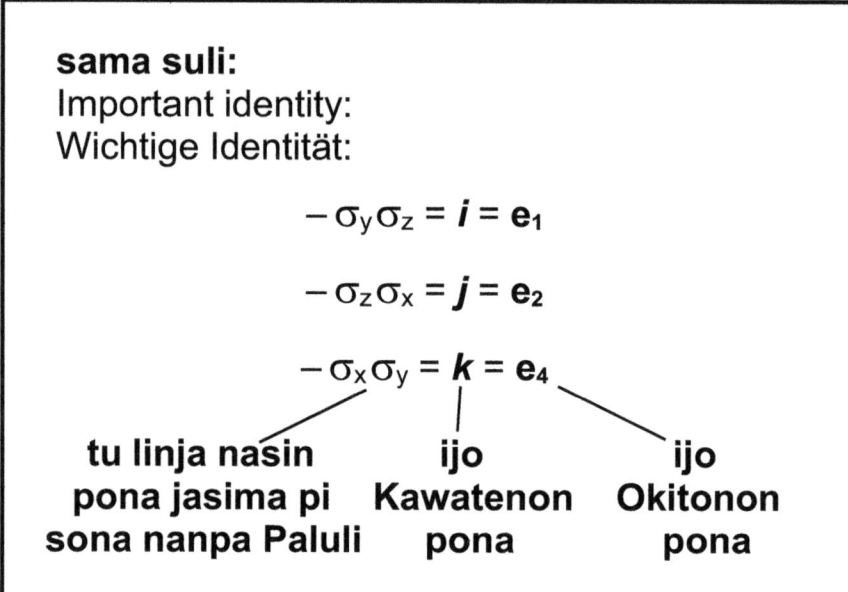

sama suli:
Important identity:
Wichtige Identität:

$$-\sigma_y\sigma_z = i = e_1$$

$$-\sigma_z\sigma_x = j = e_2$$

$$-\sigma_x\sigma_y = k = e_4$$

tu linja nasin	ijo	ijo
pona jasima pi	Kawatenon	Okitonon
sona nanpa Paluli	pona	pona

kulupu pi nanpa lon li pona mute, kin.

The multiplication of scalars is very beautiful, too.
Die Multiplikation von Skalaren ist ebenfalls sehr
schön.

$$3 \cdot 7 = 3 \clubsuit 7 = 21$$

$$5\,(-\sigma_y\sigma_z) = 5 \clubsuit (-\sigma_y\sigma_z) = -\,5\,\sigma_y\sigma_z$$

$$= 5 \clubsuit i = 5 \clubsuit e_1 = 5\,i = 5\,e_1$$

kulupu nasa ala en kulupu Okitonon li sama, tenpo sin.

The usual multiplication and the octonionic multiplication are identical again.
Die normale Multiplikation und die Oktonionen-Multiplikation sind wieder identisch.

tan ni la nanpa Kawatenon li ken kulupu e nanpa Kawatenon ante, kepeken pali lili.

Therefore quaternions can easily be multiplied by other quaternions.
Deshalb können Quaternionen einfach mit anderen Quaternionen multipliziert werden.

$(2 + 3\,i + 4\,j)\,(5 - 6\,i)$

$\qquad = (2 + 3\,e_1 + 4\,e_2)\,(5 - 6\,e_1)$

$\qquad = (2 + 3\,e_1 + 4\,e_2)\, \clubsuit\, (5 - 6\,e_1)$

$\qquad = 2 \clubsuit 5 + 3\,e_1 \clubsuit 5 + 4\,e_2 \clubsuit 5$
$\qquad\quad - 2 \clubsuit 6\,e_1 - 3\,e_1 \clubsuit 6\,e_1 - 4\,e_2 \clubsuit 6\,e_1$

$\qquad = 10 + 15\,e_1 + 20\,e_2 - 12\,e_1 + 18 + 24\,e_4$

$\qquad = 28 + 3\,e_1 + 20\,e_2 + 24\,e_4$

$\qquad = 28 + 3\,i + 20\,j + 24\,k$

kulupu ni li lon e kulupu pi selo pi linja nasin sama.

This multiplication is a multiplication of parallelograms.
Diese Multiplikation ist eine Multiplikation von Parallelogrammen.

$$(2 + 3\,i + 4\,j)\,(5 - 6\,i)$$

$$= (2 - 3\,\sigma_y\sigma_z - 4\,\sigma_z\sigma_x)\,(5 + 6\,\sigma_y\sigma_z)$$

$$= (2 - 3\,\sigma_y\sigma_z - 4\,\sigma_z\sigma_x) \clubsuit (5 + 6\,\sigma_y\sigma_z)$$

$$(2\,\sigma_z - 3\,\sigma_y + 4\,\sigma_x)\,(1\,\sigma_z) \qquad (5\,\sigma_z - 6\,\sigma_y)\,(1\,\sigma_z)$$

selo pi linja nasin   **selo pi linja nasin**
sama nanpa wan   **sama nanpa tu**
first parallelogram   second parallelogram
erstes Parallelogramm   zweites Parallelogramm

$$= (2 - 3\,\sigma_y\sigma_z - 4\,\sigma_z\sigma_x) \clubsuit (5 + 6\,\sigma_y\sigma_z)$$

$$= 28 - 3\,\sigma_y\sigma_z - 20\,\sigma_z\sigma_x - 24\,\sigma_x\sigma_y$$

$$= 28 + 3\,i + 20\,j + 24\,k$$

mi kulupu e selo pi linja nasin sama, lon ni.

We are multiplying parallelograms here.
Wir multiplizieren hier Parallelogamme.

taso nanpa Okitonon li lon e pini mute pi selo pi linja nasin sama e pini mute pi sijelo pi linja nasin sama.

But octonions are linear combinations of parallelograms and parallelepipeds.

Aber Oktonionen sind Linear-kombinationen aus Parallelo-grammen und Parallelepipeden.

mi wile sona e ni: sijelo pi linja nasin sama li kulupu, kepeken nasin seme ?

We want to learn how to multiply parallelepipeds.
Wir wollen lernen, wie Parallelepipede multipliziert werden.

taso sijelo pi linja nasin sama li lon e seme?

But what are parallelepipeds?
Aber was sind Parallelepipede?

selo pi linja nasin sama li lon e pini kulupu pi linja nasin ante tu.

Parallelograms are products of two different vectors.
Parallelogramme sind Produkte zweier verschiedener Vektoren.

sijelo pi linja nasin sama li lon e pini kulupu pi linja nasin ante tu wan.

Parallelepipeds are products of three different vectors.
Parallelepipede (Spate) sind Produkte dreier verschiedener Vektoren.

ijo pana / example / Beispiel:

$$a = \ \ 4\,\sigma_x + 1\,\sigma_y$$

$$b = \ \ 2\,\sigma_x + 5\,\sigma_y$$

$$P = a\,b = 13 + 18\,\sigma_x\sigma_y$$

$$c = -1\,\sigma_x + 3\,\sigma_z$$

$$S = a\,b\,c$$

76

(kepeken toki Tosi: Spat ... S)
sijelo pi linja nasin sama
  selo pi linja nasin sama

$$S = P\,c = a\,b\,c$$

linja nasin nanpa wan **linja nasin nanpa tu wan**
      linja nasin nanpa tu

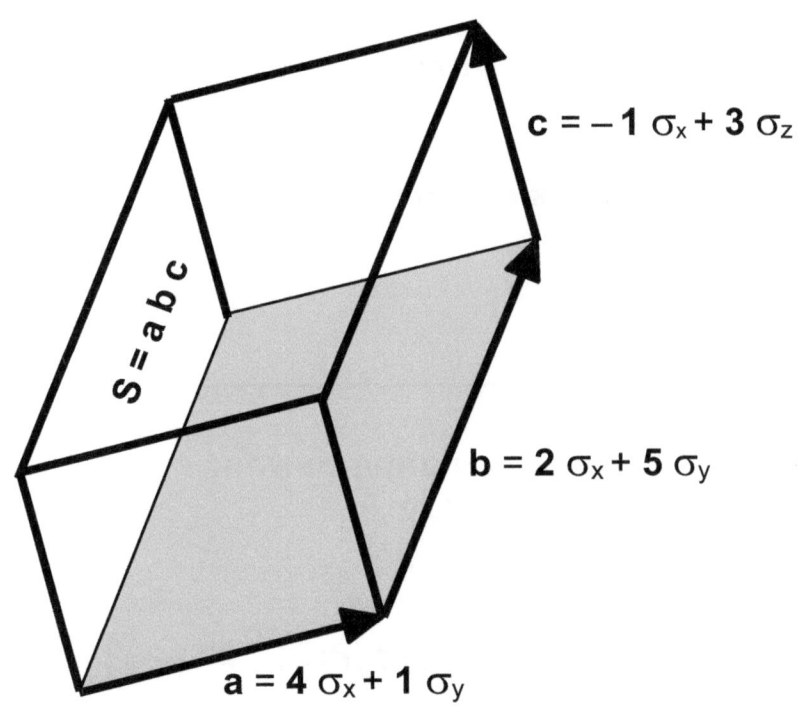

$c = -1\,\sigma_x + 3\,\sigma_z$

$S = a\,b\,c$

$b = 2\,\sigma_x + 5\,\sigma_y$

$a = 4\,\sigma_x + 1\,\sigma_y$

nanpa / computation / Rechnung:

$$S = a\,b\,c = P\,c$$

$$= (13 + 18\,\sigma_x\sigma_y)\,(-1\,\sigma_x + 3\,\sigma_z)$$

$$= -\,13\,\sigma_x - 18\,\sigma_x\sigma_y\sigma_x + 39\,\sigma_z + 54\,\sigma_x\sigma_y\sigma_z$$

$$= -\,13\,\sigma_x + 18\,\sigma_y\sigma_x^2 + 39\,\sigma_z + 54\,\sigma_x\sigma_y\sigma_z$$

$$= \underbrace{-\,13\,\sigma_x + 18\,\sigma_y + 39\,\sigma_z}_{\text{linja nasin}} + 54\underset{\text{tu wan linja nasin}}{\,\sigma_x\sigma_y\sigma_z}$$

linja nasin   **tu wan linja nasin**

tu wan linja nasin pona:

Base trivector: $I = \sigma_x\sigma_y\sigma_z$
Basis-Trivektor:

---

lawa pona nanpa tu wan:
Third basic rule:
Dritte Grundregel:

$$I^2 = (\sigma_x\sigma_y\sigma_z)^2 = -\,1$$

---

taso o kute e ni:
sijelo pi linja nasin sama ante tu li lon.

But pay attention:
Two different parallelepipeds exist.
Aber beachte:
Es gibt zwei verschiedene Parallelepipede.

ni li sijelo pi linja nasin sama sin, lon ni.

Here is the new parallelepiped.
Hier ist das neue Parallelepiped.

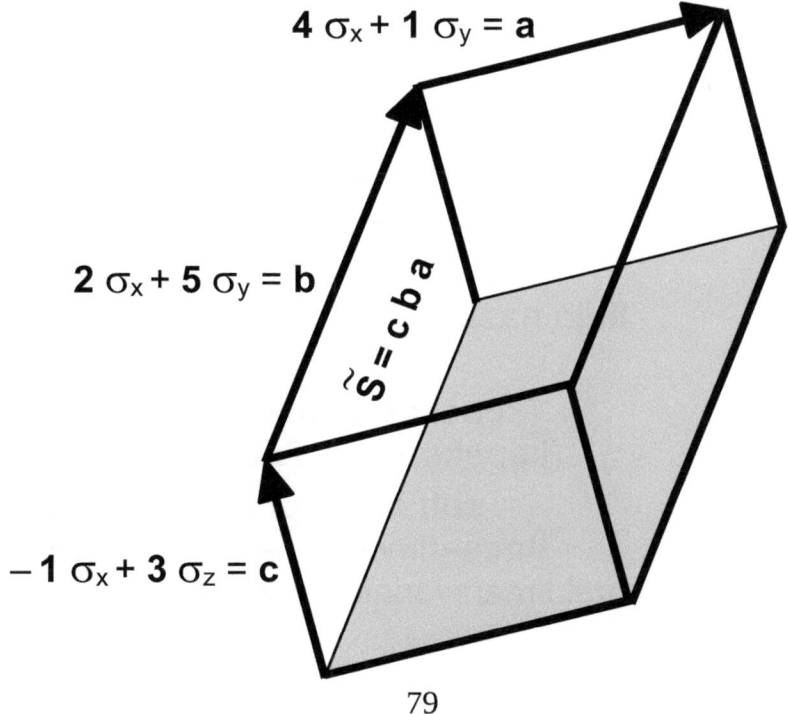

$$4\,\sigma_x + 1\,\sigma_y = a$$

$$2\,\sigma_x + 5\,\sigma_y = b$$

$$\tilde{S} = c\,b\,a$$

$$-1\,\sigma_x + 3\,\sigma_z = c$$

sijelo pi linja nasin sama sin / new parallelepiped:

neues Parallelepiped:

$$\widetilde{S} = c\,b\,a = c\,\widetilde{P}$$

$$= (-1\,\sigma_x + 3\,\sigma_z)\,(13 - 18\,\sigma_x\sigma_y)$$

$$= -13\,\sigma_x + 18\,\sigma_x{}^2\sigma_y + 39\,\sigma_z - 54\,\sigma_z\sigma_x\sigma_y$$

$$= -13\,\sigma_x + 18\,\sigma_y + 39\,\sigma_z - 54\,\sigma_x\sigma_y\sigma_z$$

$\updownarrow$ **sitelen sinpin ante**

$$S = -13\,\sigma_x + 18\,\sigma_y + 39\,\sigma_z + 54\,\sigma_x\sigma_y\sigma_z$$

mi lukin:

We see:

Wir sehen:

$$S = r + V$$

$\updownarrow$ **sitelen sinpin ante**

$$\widetilde{S} = r - V$$

linja nasin

tu wan linja nasin

(kepeken toki Inli: volume … **V)**

(kepeken toki Tosi: Volumen … **V)**

= suli sijelo pi nasin ma tu wan

three-dimensional oriented volume

dreidimensionales orientiertes Volumen

ken la linja nasin pona li kulupu.

Base vectors can be multiplied.
Basisvektoren können multipliziert werden.

ona li kulupu, lon poka pi sitelen Pano.

They multiply at the edges of the Fano diagram.
Sie multiplizieren sich entlang der Seitenkanten
des Fano-Diagramms.

sama suli:	$\sigma_x = \mathbf{e}_3$
Important identity:	$\sigma_y = \mathbf{e}_6$
Wichtige Identität:	$\sigma_z = \mathbf{e}_5$

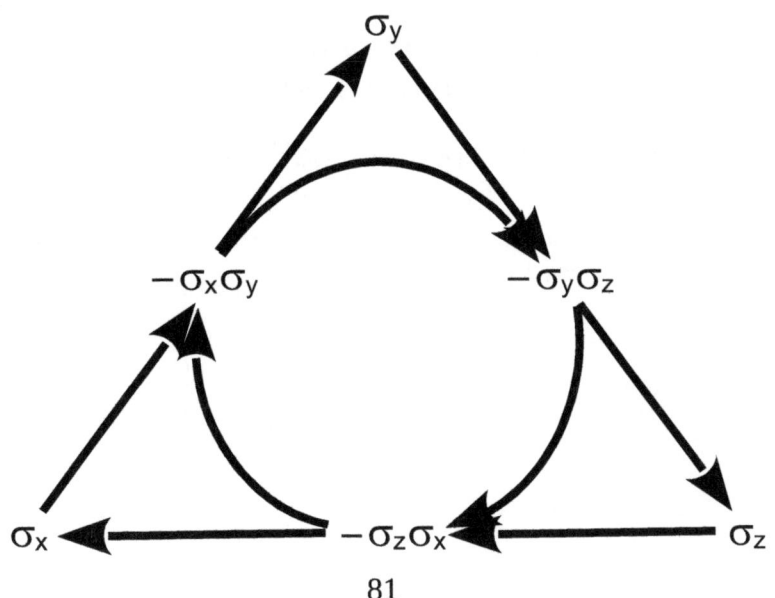

taso ni li ike. ni li lon ala !

But this is a problem. This is not correct!
Aber das ist ein Problen. Das ist nicht richtig!

$$\downarrow$$

$$\sigma_y\, \sigma_x = -\,\sigma_x\sigma_y = \sigma_y \clubsuit \sigma_x = -\,\sigma_x\sigma_y$$

$$\sigma_x\,(-\,\sigma_x\sigma_y) = -\,\sigma_y \neq \sigma_x \clubsuit (-\,\sigma_x\sigma_y) = \sigma_y$$

$$(-\,\sigma_x\sigma_y)\,\sigma_y = -\,\sigma_x \neq (-\,\sigma_x\sigma_y) \clubsuit \sigma_y = \sigma_x$$

toki pi sona nanpa nanpa wan li oko monsuta.

The first equation is an optical illusion.
Die erste Gleichung ist eine optische Täuschung.

kipisi kulupu tu li sama la toki pi sona nanpa li wile lon e ike.

If the two factors are identical, the equation will become ugly.
Wenn die beiden Faktoren identisch sind, wird die Gleichung hässlich werden.

poka tu li ante.

The two sides are different.
Die beiden Gleichungsseiten unterscheiden sich.

$$\downarrow$$

$$\sigma_x \, \sigma_x = \sigma_x^2 = 1 \neq \sigma_x \clubsuit \sigma_x = -1$$

$$\sigma_y \, \sigma_y = \sigma_y^2 = 1 \neq \sigma_y \clubsuit \sigma_y = -1$$

$$\sigma_z \, \sigma_z = \sigma_z^2 = 1 \neq \sigma_z \clubsuit \sigma_z = -1$$

sitelen jasima li lon e ike, tenpo sin.

The minus sign is a problem again.
Das Minuszeichen ist wieder ein Problem.

tan ni la mi ante e sijelo pi linja nasin sama.

Therefore we change the parallelepipeds.
Deshalb ändern wir die Parallelepipede.

mi ante e pini kulupu pi sijelo pi linja nasin sama.

We change the product of the two parallelepipeds.
Wir ändern das Produkt der beiden Parallelepipede.

pini kulupu pi selo pi linja nasin sama (P₁) en (P₂) li sama e pini kulupu Okitonon ona:

The product of two parallelograms P_1 and P_2 is identical to their octonionic product:
Das Produkt zweier Parallelogramme P_1 und P_2 ist identisch mit ihrem Oktonionenprodukt:

$$P_1 \clubsuit P_2 = P_1 \, P_2$$

taso / But / Aber:

pini kulupu pi sijelo pi linja nasin sama (S₁) en (S₂) li sama e pini kulupu Okitonon sin:

The product of two parallelepipeds S_1 and S_2 is identical to the follwowing octonionic product:
Das Produkt zweier Parallelepipede S_1 und S_2 ist identisch mit dem folgenden Oktonionenprodukt:

$$S_1 \clubsuit S_2 = - \tilde{S}_2 \, S_1$$

ni la / Thus / Also:

$$e_3 = \sigma_x \quad \Rightarrow \quad \tilde{e}_3 = \sigma_x$$

$$\Rightarrow \quad e_3 \clubsuit e_3 = \sigma_x \clubsuit \sigma_x = - \sigma_x^2 = -1$$

$$\Rightarrow \quad \textbf{lon} / \text{o.k.}$$

$$\mathbf{e}_6 = \sigma_y \quad \Rightarrow \quad \widetilde{\mathbf{e}}_6 = \sigma_y$$

$$\Rightarrow \quad \mathbf{e}_3 \clubsuit \mathbf{e}_6 = \sigma_x \clubsuit \sigma_y$$

$$= -\sigma_y\,\sigma_x = \sigma_x\sigma_y = -\mathbf{e}_4$$

$$\Rightarrow \quad \textbf{lon} \;/\; \text{o.k.}$$

kin la ni li pali, kepeken tu wan linja nasin.

This works with trivectors, too.
Das funktioniert auch mit Trivektoren.

---

sama suli:

Important identity:
Wichtige Identität: $\qquad -\sigma_x\sigma_y\sigma_z = \mathbf{e}_7$

---

ni la / Thus / Also:

$$\mathbf{e}_7 = -\sigma_x\sigma_y\sigma_z \quad \Rightarrow \quad \widetilde{\mathbf{e}}_7 = -\sigma_z\sigma_y\sigma_x = \sigma_x\sigma_y\sigma_z$$

$$\Rightarrow \quad \mathbf{e}_7 \clubsuit \mathbf{e}_7 = (-\sigma_x\sigma_y\sigma_z) \clubsuit (-\sigma_x\sigma_y\sigma_z)$$

$$= -\sigma_x\sigma_y\sigma_z\,(-\sigma_x\sigma_y\sigma_z)$$

$$= -1$$

$$\Rightarrow \quad \textbf{lon} \;/\; \text{o.k.}$$

en / and / und $\quad$ $e_3 \clubsuit e_7 = \sigma_x \clubsuit (-\sigma_x \sigma_y \sigma_z)$

$$= -\sigma_x \sigma_y \sigma_z \sigma_x$$

$$= -\sigma_y \sigma_z = e_1$$

en / and / und $\quad$ $e_7 \clubsuit e_3 = (-\sigma_x \sigma_y \sigma_z) \clubsuit \sigma_x$

$$= -\sigma_x (-\sigma_x \sigma_y \sigma_z)$$

$$= \sigma_y \sigma_z = -e_1$$

$\Rightarrow \quad$ **lon** / o.k.

tu wan linja nasin pona jasima pi sona nanpa Paluli li lon e insa pi sitelen Pano.

The negative base trivector of Pauli algebra is situated
in the middle of the Fano diagram.
Der negative Basis-Trivektor der Pauli-Algebra liegt
in der Mitte des Fano-Diagramms.

sitelen Pano li pini, tenpo ni.

Now the Fano diagram is complete.
Jetzt ist das Fano-Diagramms vollständig.

en ona li jo e nasin ma tu wan.

And it has three dimensions.
Und es besitzt drei Dimensionen.

pini pi sitelen Pano:
final Fano diagram:
endgültiges Fano-Diagramm:

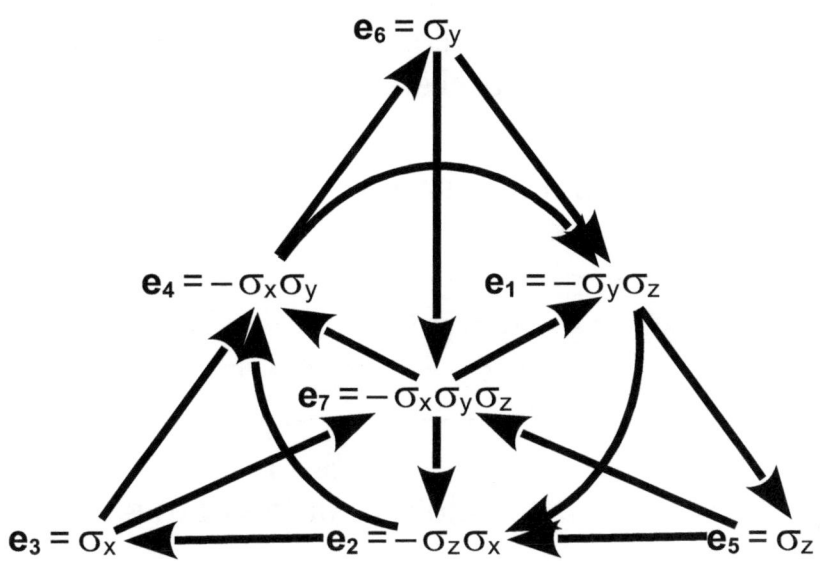

nanpa Okitonon ale li sama lukin e ni:

A complete octonion then looks like:
Ein vollständiges Oktonion sieht dann so aus:

$$a = a_0 + a_1\,\mathbf{e_1} + a_2\,\mathbf{e_2} + a_3\,\mathbf{e_3} + a_4\,\mathbf{e_4} + a_5\,\mathbf{e_5} + a_6\,\mathbf{e_6} + a_7\,\mathbf{e_7}$$

$$= a_0 + a_3\,\sigma_x + a_6\,\sigma_y + a_5\,\sigma_z$$

$$- a_4\,\sigma_x\sigma_y - a_1\,\sigma_y\sigma_z - a_2\,\sigma_z\sigma_x - a_7\,\sigma_x\sigma_y\sigma_z$$

taso mi kipisi e nanpa Okitonon, kepeken ni:
selo pi linja nasin sama
en sijelo pi linja nasin sama

But we will split the octonion into a parallelogram
and a parallelepiped.
Aber wir teilen das Oktonion in ein Parallelogramm
und ein Parallelepiped auf.

selo pi linja nasin sama
$$P = k + A$$

$$a = a_0 - a_4\,\sigma_x\sigma_y - a_1\,\sigma_y\sigma_z - a_2\,\sigma_z\sigma_x$$
$$+ a_3\,\sigma_x + a_6\,\sigma_y + a_5\,\sigma_z - a_7\,\sigma_x\sigma_y\sigma_z$$

S = r + V
sijelo pi linja nasin sama

$$\Rightarrow \qquad a = P + S$$

mi kama jo e nanpa Okitonon tu, tenpo ni.

Now we will take two octonions.
Jetzt nehmen wir zwei Oktonionen.

$$\Rightarrow \qquad a_1 = P_1 + S_1$$
$$a_2 = P_2 + S_2$$

88

en mi kulupu e ona, kepeken kulupu Okitonon.

And we multiply them with the octonion multiplication.
Und wir multiplizieren sie mit Hilfe der Oktonionen-Multiplikation.

$$a_1 \clubsuit a_2 = (P_1 + S_1) \clubsuit (P_2 + S_2)$$

$$= P_1 \clubsuit P_2 + P_1 \clubsuit S_2 + S_1 \clubsuit P_2 + S_1 \clubsuit S_2$$

$$= P_1 P_2 + \underbrace{\quad ? \quad + \quad ? \quad} - \tilde{S}_2 S_1$$

kipisi insa li weka.

The terms in the middle are missing.
Die mittleren Terme fehlen.

selo pi linja nasin sama li ken kulupu e sijelo pi linja nasin sama, kepeken kulupu Okitonon:

A parallelogram can be multiplied by a parallelepiped by the octonionic multiplication in the following way:
Ein Parallelogramm kann mit Hilfe der Oktonionen-Multiplikation folgendermaßen mit einem Parallelepiped multipliziert werden:

$$P_1 \clubsuit S_2 = S_2 P_1$$

ni la / Thus / Also:

$$e_4 = -\sigma_x \sigma_y$$

$$e_3 = \sigma_x \qquad \Rightarrow \qquad e_4 \clubsuit e_3 = (-\sigma_x \sigma_y) \clubsuit \sigma_x = \sigma_x (-\sigma_x \sigma_y)$$
$$= -\sigma_y = -e_6$$

$$e_6 = \sigma_y \qquad \Rightarrow \qquad e_4 \clubsuit e_6 = (-\sigma_x \sigma_y) \clubsuit \sigma_y = \sigma_y (-\sigma_x \sigma_y)$$
$$= \sigma_x = e_3$$

$$e_5 = \sigma_z \qquad \Rightarrow \qquad e_4 \clubsuit e_5 = (-\sigma_x \sigma_y) \clubsuit \sigma_z = \sigma_z (-\sigma_x \sigma_y)$$
$$= -\sigma_x \sigma_y \sigma_z = e_7$$

$$\Rightarrow \qquad \textbf{lon} / \text{o.k.}$$

en / and / und

$$e_7 = -\sigma_x \sigma_y \sigma_z \qquad \Rightarrow \qquad e_4 \clubsuit e_7 = (-\sigma_x \sigma_y) \clubsuit (-\sigma_x \sigma_y \sigma_z)$$
$$= \sigma_x \sigma_y \sigma_z \, \sigma_x \sigma_y$$
$$= -\sigma_z = -e_5$$

$$\Rightarrow \qquad \textbf{lon} / \text{o.k.}$$

en lon a / and of course / und natürlich:

$$42 \clubsuit e_7 = 42 \clubsuit (-\sigma_x \sigma_y \sigma_z)$$
$$= -\sigma_x \sigma_y \sigma_z \, (42) = -42 \, \sigma_x \sigma_y \sigma_z = 42 \, e_7$$

$$\Rightarrow \qquad \textbf{lon} / \text{o.k.}$$

sijelo pi linja nasin sama li ken kulupu e selo pi linja nasin sama, kepeken kulupu Okitonon:

A parallelepiped can be multiplied by a parallelogram by the octonionic multiplication in the following way:
Ein Parallelepiped kann mit Hilfe der Oktonionen-Multiplikation folgendermaßen mit einem Parallelogramm multipliziert werden:

$$S_1 \clubsuit P_2 = S_1 \widetilde{P}_2$$

ni la / Thus / Also:

$e_3 = \sigma_x$

$e_4 = -\sigma_x\sigma_y \quad \Rightarrow \quad e_3 \clubsuit e_4 = \sigma_x \clubsuit (-\sigma_x\sigma_y) = \sigma_x(-\sigma_y\sigma_x)$

$$= \sigma_y = e_6$$

$e_1 = -\sigma_y\sigma_z \quad \Rightarrow \quad e_3 \clubsuit e_1 = \sigma_x \clubsuit (-\sigma_y\sigma_z) = \sigma_x(-\sigma_z\sigma_y)$

$$= \sigma_x\sigma_y\sigma_z = -e_7$$

$$\Rightarrow \quad \textbf{lon} / \text{o.k.}$$

en / and / und

$e_7 = -\sigma_x\sigma_y\sigma_z$

$e_4 = -\sigma_x\sigma_y \quad \Rightarrow \quad e_7 \clubsuit e_4 = (-\sigma_x\sigma_y\sigma_z) \clubsuit (-\sigma_x\sigma_y)$

$$= (-\sigma_x\sigma_y\sigma_z)(-\sigma_y\sigma_x)$$

$$= \sigma_x \sigma_y \sigma_z \, \sigma_y \sigma_x$$

$$= \sigma_z = \mathbf{e_5}$$

$$\Rightarrow \quad \textbf{lon} \,/\, \text{o.k.}$$

en lon a / and of course / und natürlich

$$\mathbf{e_7} \clubsuit \mathbf{42} = (-\,\sigma_x \sigma_y \sigma_z) \clubsuit \mathbf{42}$$

$$= -\,\sigma_x \sigma_y \sigma_z \,(\mathbf{42}) = -\,\mathbf{42}\,\sigma_x \sigma_y \sigma_z = \mathbf{42}\,\mathbf{e_7}$$

$$\Rightarrow \quad \textbf{lon} \,/\, \text{o.k.}$$

mi pakala e kulupu Okitonon.

We destroy octonionic multiplication.
Wir zerstören die Oktonionen-Multiplikation.

mi ante e ona.

We replace it.
Wir ersetzen sie.

mi ante e ona, kepeken sona nanpa Paluli.

We replace it by Pauli algebra.
Wir ersetzen sie durch die Pauli-Algebra.

ni li lon e supa lape pi jan Lasenpi.

This is the Lasenby embedding.
Das ist die Lasenby-Einbettung.

$$a_1 \clubsuit a_2 = (P_1 + S_1) \clubsuit (P_2 + S_2)$$

$$= P_1 \clubsuit P_2 + P_1 \clubsuit S_2 + S_1 \clubsuit P_2 + S_1 \clubsuit S_2$$

$$= P_1 P_2 + S_2 P_1 + S_1 \tilde{P}_2 - \tilde{S}_2 S_1$$

toki pi sona nanpa ni li suli.

This equation is important.
Diese Gleichung ist wichtig.

toki pi sona nanpa ni li suli mute.

This equation is very important.
Diese Gleichung ist sehr wichtig.

sona nanpa ni li pana sona e nasin insa.

This mathematics explains the standard model.
Diese Mathematik erklärt das Standardmodell.

o lukin e ni / see / siehe:

Anthony Lasenby: Geometric Algebra, Octonions and the standard model. OSMU Talk 17 at Ocotber 27th, 2023, URL: https://www.youtube.com/watch?v=0m__fhtkMzg [07. Nov. 2023].

Anthony Lasenby: Some recent results for *SU(3)* and Octonions within the Geometric Algebra approach to the fundamental forces of nature. In: Mathematical Methoths in the Applied Sciences, Vol. 47, No. 3 (2024), pp. 1471 – 1491, and: arXiv:2202.06733v1, URL: https://arxiv.org/pdf/2202.06733 [09. Feb. 2022].

Eckhard Hitzer: Extending Lasenby's embedding of octonions in space-time algebra *Cl(1; 3)*, to all three- and four-dimensional Clifford geometric algebras *Cl(p; q)*, $n = p + q = 3; 4$. In: Mathematical Methoths in the Applied Sciences, Vol. 47, No. 3 (2024), pp. 1401 – 1424.

Martin Erik Horn: Tetrapodenrechnerei – Tetrapod Computations. Auch mit tetrapodischen Quaternionen und tetrapodischen Oktonionen – Including tetrapodic quaternions and tetrapodic octonions. BoD, Norderstedt 2025 (Kap. 17: Oktonionen / Chap. 17: Octonions).

mi kepeken e supa lape pi jan Lasenpi, tawa ni: mi wile nanpa ijo pana sin ala, tenpo sin.

We will use the Lasenby embedding to solve an old example again.
Wir benutzen die Lasenby-Einbettung, um ein altes Beispiel erneut durchzurechnen.

ijo pana (lipu nanpa 33):
example (page 33):
Beispiel (Seite 33):

$$a = 12 + 5\,e_1 = 12 - 5\,\sigma_y\sigma_z$$

$$\Rightarrow \quad P_1 = 12 - 5\,\sigma_y\sigma_z \quad S_1 = 0$$

$$b = 3\,e_2 + 4\,e_3 = 4\,\sigma_x - 3\,\sigma_z\sigma_x$$

$$\Rightarrow \quad P_2 = -3\,\sigma_z\sigma_x \qquad S_2 = 4\,\sigma_x$$

$$c = 3 + 2\,e_6 + 6\,e_7 = 3 + 2\,\sigma_y - 6\,\sigma_x\sigma_y\sigma_z$$

$$\Rightarrow \quad P_4 = 3 \qquad\qquad S_4 = 2\,\sigma_y - 6\,\sigma_x\sigma_y\sigma_z$$

$$(a \clubsuit b) \clubsuit c = ? \qquad a \clubsuit (b \clubsuit c) = ?$$

o kepeken sona nanpa Paluli !

Use Pauli algebra!
Verwende die Pauli-Algebra!

nanpa pi toki pi sona nanpa nanpa wan, kepeken sona nanpa Paluli:

Computation of the first equation with Pauli algebra:
Berechnung der ersten Gleichung mit Hilfe der Pauli-Algebra:

$$a \clubsuit b = P_1 P_2 + S_2 P_1 + S_1 \tilde{P}_2 - \tilde{S}_2 S_1$$

$$P_1 P_2 = (12 - 5\, \sigma_y \sigma_z)(-3\, \sigma_z \sigma_x)$$

$$= -36\, \sigma_z \sigma_x - 15\, \sigma_x \sigma_y$$

$$S_2 P_1 = 4\, \sigma_x (12 - 5\, \sigma_y \sigma_z)$$

$$= 48\, \sigma_x - 20\, \sigma_x \sigma_y \sigma_z$$

$$S_1 \tilde{P}_2 = \tilde{S}_2 S_1 = 0$$

$$a \clubsuit b = 48\, \sigma_x - 15\, \sigma_x \sigma_y - 36\, \sigma_z \sigma_x - 20\, \sigma_x \sigma_y \sigma_z$$

$$\Rightarrow \qquad P_3 = -15\, \sigma_x \sigma_y - 36\, \sigma_z \sigma_x$$

$$S_3 = 48\, \sigma_x - 20\, \sigma_x \sigma_y \sigma_z$$

$$(a \clubsuit b) \clubsuit c = P_3 P_4 + S_4 P_3 + S_3 \tilde{P}_4 - \tilde{S}_4 S_3$$

$$P_3 P_4 = (-15\, \sigma_x \sigma_y - 36\, \sigma_z \sigma_x)(3)$$

$$= -45\, \sigma_x \sigma_y - 108\, \sigma_z \sigma_x$$

$$S_4\, P_3 = (2\,\sigma_y - 6\,\sigma_x\sigma_y\sigma_z)\,(-15\,\sigma_x\sigma_y - 36\,\sigma_z\sigma_x)$$

$$= 30\,\sigma_x - 216\,\sigma_y - 90\,\sigma_z - 72\,\sigma_x\sigma_y\sigma_z$$

$$S_3\,\widetilde{P}_4 = (48\,\sigma_x - 20\,\sigma_x\sigma_y\sigma_z)\,(3)$$

$$= 144\,\sigma_x - 60\,\sigma_x\sigma_y\sigma_z$$

$$\widetilde{S}_4\,S_3 = (2\,\sigma_y + 6\,\sigma_x\sigma_y\sigma_z)\,(48\,\sigma_x - 20\,\sigma_x\sigma_y\sigma_z)$$

$$= 120 - 96\,\sigma_x\sigma_y + 288\,\sigma_y\sigma_z - 40\,\sigma_z\sigma_x$$

$$-\,\widetilde{S}_4\,S_3 = -120 + 96\,\sigma_x\sigma_y - 288\,\sigma_y\sigma_z + 40\,\sigma_z\sigma_x$$

$$(a \clubsuit b) \clubsuit c = -120 + 174\,\sigma_x - 216\,\sigma_y - 90\,\sigma_z + 51\,\sigma_x\sigma_y$$
$$-\,288\,\sigma_y\sigma_z - 68\,\sigma_z\sigma_x - 132\,\sigma_x\sigma_y\sigma_z$$

$$= -120 + 288\,e_1 + 68\,e_2 + 174\,e_3 - 51\,e_4$$
$$-\,90\,e_5 - 216\,e_6 + 132\,e_7$$

pini pona li sama e pini pona sin ala, insa lipu 33.

The result is identical to the old result of page 33.
Das Ergebnis stimmt mit dem alten Ergebnis von Seite 33 überein.

pini pona ni li lon.

This result is correct.
Dieses Ergebnis ist richtig.

nanpa pi toki pi sona nanpa nanpa tu, kepeken sona nanpa Paluli:

Computation of the second equation with Pauli algebra:

Berechnung der zweiten Gleichung mit Hilfe der Pauli-Algebra:

$$b \clubsuit c = P_2 P_4 + S_4 P_2 + S_2 \tilde{P}_4 - \tilde{S}_4 S_2$$

$$P_2 P_4 = (-3\,\sigma_z\sigma_x)(3) = -9\,\sigma_z\sigma_x$$

$$S_4 P_2 = (2\,\sigma_y - 6\,\sigma_x\sigma_y\sigma_z)(-3\,\sigma_z\sigma_x)$$

$$= -18\,\sigma_y - 6\,\sigma_x\sigma_y\sigma_z$$

$$S_2 \tilde{P}_4 = (4\,\sigma_x)(3) = 12\,\sigma_x$$

$$\tilde{S}_4 S_2 = (2\,\sigma_y + 6\,\sigma_x\sigma_y\sigma_z)(4\,\sigma_x)$$

$$= -8\,\sigma_x\sigma_y + 24\,\sigma_y\sigma_z$$

$$-\tilde{S}_4 S_2 = 8\,\sigma_x\sigma_y - 24\,\sigma_y\sigma_z$$

$$b \clubsuit c = 12\,\sigma_x - 18\,\sigma_y + 8\,\sigma_x\sigma_y - 24\,\sigma_y\sigma_z$$
$$-9\,\sigma_z\sigma_x - 6\,\sigma_x\sigma_y\sigma_z$$

$$\Rightarrow \quad P_5 = 8\,\sigma_x\sigma_y - 24\,\sigma_y\sigma_z - 9\,\sigma_z\sigma_x$$

$$S_5 = 12\,\sigma_x - 18\,\sigma_y - 6\,\sigma_x\sigma_y\sigma_z$$

$$\mathbf{a} \clubsuit (\mathbf{b} \clubsuit \mathbf{c}) = \mathbf{P_1\,P_5} + \mathbf{S_5\,P_1} + \mathbf{S_1\,\widetilde{P}_5} - \mathbf{\widetilde{S}_5\,S_1}$$

$$\mathbf{P_1\,P_5} = (12 - 5\,\sigma_y\sigma_z)\,(8\,\sigma_x\sigma_y - 24\,\sigma_y\sigma_z - 9\,\sigma_z\sigma_x)$$

$$= -120 + 51\,\sigma_x\sigma_y - 288\,\sigma_y\sigma_z - 148\,\sigma_z\sigma_x$$

$$\mathbf{S_5\,P_1} = (12\,\sigma_x - 18\,\sigma_y - 6\,\sigma_x\sigma_y\sigma_z)\,(12 - 5\,\sigma_y\sigma_z)$$

$$= 114\,\sigma_x - 216\,\sigma_y + 90\,\sigma_z - 132\,\sigma_x\sigma_y\sigma_z$$

$$\mathbf{S_1\,\widetilde{P}_5} = \mathbf{\widetilde{S}_5\,S_1} = 0$$

$$\mathbf{a} \clubsuit (\mathbf{b} \clubsuit \mathbf{c}) = -120 + 114\,\sigma_x - 216\,\sigma_y + 90\,\sigma_z + 51\,\sigma_x\sigma_y$$
$$- 288\,\sigma_y\sigma_z - 148\,\sigma_z\sigma_x - 132\,\sigma_x\sigma_y\sigma_z$$

$$= -120 + 288\,e_1 + 148\,e_2 + 114\,e_3 - 51\,e_4$$
$$+ 90\,e_5 - 216\,e_6 + 132\,e_7$$

pini pona li sama e pini pona sin ala, insa lipu 35.

The result is identical to the old result of page 35.
Das Ergebnis stimmt mit dem alten Ergebnis von
Seite 35 überein.

pini pona ni li lon.

This result is correct.
Dieses Ergebnis ist richtig.

supa lape pi jan Lasenpi li pali.

The Lasenby embedding works.
Die Lasenby-Einbettung funktioniert.

ni li pona lukin.

This is beautiful.
Das ist wunderschön.

**tan ni la mi awen e sona nanpa Kijaki.
ona li pona lukin, kin.**

Therefore we are waiting for Dirac algebra.
It is beautiful too.
Deshalb warten wir auf die Dirac-Algebra.
Sie ist ebenfalls wunderschön.

**mi ken sitelen e supa lape pi jan Lasenpi,
kepeken sona nanpa Kijaki.**

We are able to write the Lasenby embedding
with Dirac algebra.
Wir können die Lasenby-Einbettung mit Hilfe der
Dirac-Algebra schreiben.

sona nanpa Kijaki li wile kama.

Dirac algebra will come.
Die Dirac-Algebra wird kommen.

lipu namako / Attachment / Anhang

nimi / Names / Namen

The persons mentioned in this book are called by the following tokiponazed names:

Die in diesem Buch erwähnten Personen werden mit folgenden tokiponaisierten Namen bezeichnet:

jan Kaku	Michio Kaku
jan Kasaman	Hermann Grassmann
jan Kijaki	Paul A. M. Dirac
jan Konwe nanpa tu	John H. Conway
jan Konwe nanpa wan	Arthur W. Conway
jan Lasenpi	Anthony Lasenby
jan Paluli	Wolfgang Pauli
jan Pano	Gino Fano
jan Pu	Clumsy Foo
jan Puwe	Cohl Furey
jan Soje	qam Soy'

kon nimi / Vocabulary / Vokabeln

As the semantic space of Toki Pona words is immense, some special meanings of the words used in this book are given in the following.

Da der semantische Raum von Worten in Toki Pona enorm ist, werden im Folgenden einige spezielle Bedeutungen der Worte, die in diesem Buch genutzt werden, vorgestellt.

insa sike	circular, in a circle	zyklisch, kreisförmig
ijo jasima	negative quantity	negative Größe
ijo lon	real quantity	reelle Größe

ijo lon pona	real base unit	reelle Basisgröße
ijo musi	complex quantity	komplexe Größe
ijo musi mute	hypercomplex quantity	hyperkomplexe Größe
ijo nasa	imaginary quantity	imaginäre Größe
ijo nasa pona	imaginary base unit	imaginäre Basisgröße
ijo pana	example	Beispiel
jasima	opposite, minus, reflection, negative	entgegengesetzt, minus, Reflexion, negativ
kipisi	part, term, division to split, to divide,	Teil, Term, Division, teilen, dividieren,
kipisi kulupu	factor	Faktor
kipisi lon	real part, real term	reeler Teil, reeller Term
kipisi nasa	imaginary part, imaginary term	imaginärer Teil, imaginärer Term
kulupu	to form a group, to multiply, times, multiplication	gruppieren, multiplizieren, mal, Multiplikation
lili	small, short, to decrease, to subtract, subtraction	klein, kurz, verringern, subtrahieren, Subtraktion
linja nasin	line with direction, vector	Linie mit Richtung, Vektor
linja nasin pona	unit vector, base vector	Einheitsvektor, Basisvektor
mute	several, to increase to add to, addition	mehrere, erhöhen addieren zu, Addition
nanpa	number, to compute, to calculate, calculation, computation	Zahl, rechnen, berechnen, Rechnung, Berechnung
nanpa Kawatenon	quaternion	Quaternion
nanpa lon	real number, scalar	reelle Zahl, Skalar
nanpa musi	complex number	komplexe Zahl
nanpa nasa	imaginary number	imaginäre Zahl
nanpa Okitonon	octonion	Oktonion

nasin	way, direction, orientation, order	Weg, Richtung, Orientierung, Reihenfolge
nasin jasima pi ilo tenpo	anti-clockwise orientation	Orientierung entgegen dem Uhrzeigersinn
nasin ma	direction of space, dimension	Richtung der Welt, Dimension
nasin pi ilo tenpo	clockwise orientation	Orientierung im Uhrzeigersinn
nasin sama	parallel	parallel
pali tonsi musi	complex conjugation	komplexe Konjugation
pini kipisi	end ef splitting, decomposition, quotient	Ende der Teilung, Zerlegung, Quotient
pini kulupu	end of multiplication, product	Ende der Multiplikation, Produkt
pini lili	end of subtraction, difference	Ende der Subtraktion, Differenz
pini mute	end of addition, sum, linear combination	Ende der Addition, Summe, Linearkombination
pini pona	result, solution	Ergebnis, Lösung
poka sike	(round) bracket	(runde) Klammer
selo pi linja nasin sama	form of parallel lines, parallelogram	Form aus parallelen Linien, Parallelogramm
selo pi linja tu wan	form of three lines, triangle	Form aus drei Linien, Dreieck
sijelo	body, structure	Körper, Struktur
sijelo musi	complex structure	komplexe Struktur
sijelo pi linja nasin sama	body of parallel lines, parallelepiped	Parallelepiped, Spat
sijelo sona nanpa	mathematical structure	mathematische Struktur
sitelen	symbol, sign to depict, to write	Symbol, Zeichen darstellen, schreiben
sitelen jasima	minus sign	Minuszeichen
sitelen lili	index, indices	Index, Indizes

sitelen Pano	Fano diagram	Fano-Diagramm
sitelen sinpin	algebraic sign	Vorzeichen, Rechenzeichen
sona nanpa	mathematics	Mathematik
sona nanpa Kawatenon	quaternion algebra	Quaternionenalgebra
sona nanpa Okitonon	octonion algebra	Oktonionenalgebra
sona nanpa Paluli	Pauli algebra	Pauli-Algebra
sona nasin	philosophy	Philosophie
sona pini	conclusion	Schlussfolgerung
sona tawa	physics	Physik
suli	large, long, magnitude, normed, important	groß, lang, Betrag, normiert, wichtig
toki pi sona nanpa	equation	Gleichung
tomo sona	university	Universität
tomo sona Kenpite	University of Cambridge	Universität Cambridge
tomo sona Pinseton	University of Princeton	Universität Princeton
tu linja nasin	bivector	Bivektor
tu linja nasin pona	unit bivector, base bivector	Einheitsbivektor, Basisbivektor
tu wan linja nasin	trivector	Trivektor
tu wan linja nasin pona	unit trivector, base trivector	Einheitstrivektor, Basistrivektor
wan jasima	minus one, -1	minus Eins, -1
wan pona	plus one, $+1$	plus Eins, $+1$
weka poka sike	without brackets, associative	ohne Klammern, assoziativ